The Mechanical Properties of Wood

Including a Discussion of the Factors Affecting the Mechanical Properties, and Methods of Timber Testing

by

Samuel J. Record

Photomicrograph of a small block of western hemlock. At the top is the cross section showing to the right the late wood of one season's growth. to the left the early wood of the next season. The other two sections are longitudinal and show the fibrous character of the wood. To the left is the radial section with three rays crossing it. To the right is the tangential section upon which the rays appear as vertical rows of beads. × 35.

PREFACE

This book was written primarily for students of forestry to whom a knowledge of the technical properties of wood is essential. The mechanics involved is reduced to the simplest terms and without reference to higher mathematics, with which the students rarely are familiar. The intention throughout has been to avoid all unnecessarily technical language and descriptions, thereby making the subject-matter readily available to every one interested in wood.

Part I is devoted to a discussion of the mechanical properties of wood—the relation of wood material to stresses and strains. Much of the subject-matter is merely elementary mechanics of materials in general, though written with reference to wood in particular. Numerous tables are included, showing the various strength values of many of the more important American woods.

Part II deals with the factors affecting the mechanical properties of wood. This is a subject of interest to all who are concerned in the rational use of wood, and to the forester it also, by retrospection, suggests ways and means of regulating his forest product through control of the conditions of production. Attempt has been made, in the light of all data at hand, to answer many moot questions, such as the effect on the quality of wood of rate of growth, season of cutting, heartwood and sapwood, locality of growth, weight, water content, steaming, and defects.

Part III describes methods of timber testing. They are for the most part those followed by the U. S. Forest Service. In schools equipped with the necessary machinery the instructions will serve to direct the tests; in others a study of the text with reference to the illustrations should give an adequate conception of the methods employed in this most important line of research.

The appendix contains a copy of the working plan followed by the U. S. Forest Service in the extensive investigations covering the mechanical properties of the woods grown in the United

States. It contains many valuable suggestions for the independent investigator. In addition four tables of strength values for structural timbers, both green and air-seasoned, are included. The relation of the stresses developed in different structural forms to those developed in the small clear specimens is given.

In the bibliography attempt was made to list all of the important publications and articles on the mechanical properties of wood, and timber testing. While admittedly incomplete, it should prove of assistance to the student who desires a fuller knowledge of the subject than is presented here.

The writer is indebted to the U. S. Forest Service for nearly all of his tables and photographs as well as many of the data upon which the book is based, since only the Government is able to conduct the extensive investigations essential to a thorough understanding of the subject. More than eighty thousand tests have been made at the Madison laboratory alone, and the work is far from completion.

The writer also acknowledges his indebtedness to Mr. Emanuel Fritz, M.E., M.F., for many helpful suggestions in the preparation of Part I; and especially to Mr. Harry Donald Tiemann, M.E., M.F., engineer in charge of Timber Physics at the Government Forest Products Laboratory, Madison, Wisconsin, for careful revision of the entire manuscript.

SAMUEL J. RECORD.

YALE FOREST SCHOOL, *July* 1, 1914.

CONTENTS

PAGE

PREFACE V

PART I

THE MECHANICAL PROPERTIES OF WOOD
Introduction 1
Fundamental considerations and definitions . . . 2
Tensile strength 7
Compressive or crushing strength . . 9
Shearing strength 19
Transverse or bending strength: Beams . . . 22
Toughness: Torsion 37
Hardness 39
Cleavability . 40

PART II

FACTORS AFFECTING THE MECHANICAL PROPERTIES OF WOOD
Introduction 43
Rate of growth 43
Heartwood and sapwood 50
Weight, density, and specific gravity 54
Color 58
Cross grain 59
Knots 61
Frost splits 62
Shakes, galls, pitch pockets 64
Insect injuries 66
Marine wood-borer injuries 67
Fungous injuries 68
Parasitic plant injuries 70
Locality of growth 70
Season of cutting 73
Water content 75
Temperature 84
Preservatives 86

PART III

Timber Testing   PAGE
  Working plan . . .   88
  Forms of material tested   88
  Size of test specimens . .   89
  Moisture determination . . .   90
  Machine for static tests .   90
  Speed of testing machine   92
  Bending large beams . . .   94
  Bending small beams . . .   99
  Endwise compression   102
  Compression across the grain . . .   104
  Shear along the grain   107
  Impact test   110
  Hardness test: Abrasion and indentation   114
  Cleavage test . .   118
  Tension test parallel to the grain   118
  Tension test at right angles to the grain   120
  Torsion test   122
  Special tests   123
  Spike pulling test   123
  Packing boxes   124
  Vehicle and implement woods   124
  Cross-arms   124
  Other tests   125

Appendix
  Sample working plan of United States Forest Service . . . 127
  Strength values for structural timbers 138

Bibliography 145
  Part I: Some general works on mechanics, materials of construction,
    and testing of materials 147
  Part II: Publications and articles on the mechanical properties of
    wood, and timber testing 148
  Part III: Publications of the United States Government on the
    mechanical properties of wood, and timber testing . . 157

Index 161

ILLUSTRATIONS

FIG.   PAGE

Photomicrograph of a small block of western hemlock   *Frontispiece*

1.—Stress-strain diagrams of two longleaf pine beams   4

2.—Compression across the grain   10

3.—Side view of failures in compression across the grain . . .   10

4.—End view of failures in compression across the grain   10

5.—Testing a buggy-spoke in endwise compression   11

6.—Unequal distribution of stress in a long column due to lateral
   bending   12

7.—Endwise compression of a short column . . .   15

8.—Failures of a short column of green spruce   17

9.—Failures of short columns of dry chestnut . . .   18

10.—Example of shear along the grain   19

11.—Failures of test specimens in shear along the grain . .   19

12.—Horizontal shear in a beam   21

13.—Oblique shear in a short column   21

14.—Failure of a short column by oblique shear   21

15.—Diagram of a simple beam   23

16.—Three common forms of beams—(1) simple, (2) cantilever, (3)
   continuous   24

17.—Characteristic failures of simple beams   35

18.—Failure of a large beam by horizontal shear   37

19.—Torsion of a shaft   38

20.—Effect of torsion on different grades of hickory   39

21.—Cleavage of highly elastic wood . . .   41

22.—Cross-sections of white ash, red gum, and eastern hemlock . . .   45

23.—Cross-section of longleaf pine   46

24.—Relation of the moisture content to the various strength values of
   spruce   77

25.—Cross-section of the wood of western larch showing fissures in the
   thick-walled cells of the late wood   79

26.—Progress of drying throughout the length of a chestnut beam   79

27.—Excessive season checking . . .   81

28.—Control of season checking by the use of S-irons . . .   83

29.—Static bending test on a large beam   95

30.—Two methods of loading a beam   97

31.—Static bending test on a small beam   99

32.—Sample log sheet, giving full details of a transverse bending test on
   a small pine beam   101

FIG.   PAGE
33.—Endwise compression test   103
34.—Sample log sheet of an endwise compression test on a short pine
        column   105
35.—Compression across the grain   106
36.—Vertical section of shearing tool   .   .   107
37.—Front view of shearing tool   .   .   108
38.—Two forms of shear test specimens   108
39.—Making a shearing test   109
40.—Impact testing machine   111
41.—Drum record of impact bending test   113
42.—Abrasion machine for testing the wearing qualities of woods   .   115
43.—Design of tool for testing the hardness of woods by indentation   .   . 116
44.—Design of tool for cleavage test   117
45.—Design of cleavage test specimen   118
46.—Designs of tension test specimens used in United States   .   119
47.—Design of tension test specimen used in New South Wales   .   120
48.—Design of tool and specimen for testing tension at right angles to the
        grain   121
49.—Making a torsion test on hickory   .   .   . 122
50.—Method of cutting and marking test specimens   129
51.—Diagram of specific gravity apparatus   137

TABLES

NO.		PAGE
I.—Comparative strength of iron, steel, and wood		7
II.—Ratio of strength of wood in tension and in compression . .		8
III.—Right-angled tensile strength of small clear pieces of 25 woods in green condition		9
IV.—Results of compression tests across the grain on 51 woods in green condition, and comparison with white oak		13
V.—Relation of fibre stress at elastic limit in bending to the crushing strength of blocks cut therefrom in pounds per square inch .		14
VI.—Results of endwise compression tests on small clear pieces of 40 woods in green condition		16
VII.—Shearing strength along the grain of small clear pieces of 41 woods in green condition		20
VIII.—Shearing strength across the grain of various American woods .		22
IX.—Results of static bending tests on small clear beams of 49 woods in green condition		27
X.—Results of impact bending tests on small clear beams of 34 woods in green condition		32
XI.—Manner of first failure of large beams		36
XII.—Hardness of 32 woods in green condition, as indicated by the load required to imbed a 0.444-inch steel ball to one-half its diameter		40
XIII.—Cleavage strength of small clear pieces of 32 woods in green condition		42
XIV.—Specific gravity, and shrinkage of 51 American woods . . .		56
XV.—Effect of drying on the mechanical properties of wood, shown in ratio of increase due to reducing moisture content from the green condition to kiln-dry		76
XVI.—Effect of steaming on the strength of green loblolly pine . .		85
XVII.—Speed-strength moduli, and relative increase in strength at rates of fibre strain increasing in geometric ratio		93
XVIII.—Results of bending tests on green structural timbers		140
XIX.—Results of compression and shear tests on green structural timbers		141
XX.—Results of bending tests on air-seasoned structural timbers . .		142
XXI.—Results of compression and shear tests on air-seasoned structural timbers		143
XXII.—Working unit stresses for structural timber expressed in pounds per square inch		144

PART I

THE MECHANICAL PROPERTIES OF WOOD

INTRODUCTION

THE mechanical properties of wood are its fitness and ability to resist applied or external forces. By external force is meant any force outside of a given piece of material which tends to deform it in any manner. It is largely such properties that determine the use of wood for structural and building purposes and innumerable other uses of which furniture, vehicles, implements, and tool handles are a few common examples.

Knowledge of these properties is obtained through experimentation either in the employment of the wood in practice or by means of special testing apparatus in the laboratory. Owing to the wide range of variation in wood it is necessary that a great number of tests be made and that so far as possible all disturbing factors be eliminated. For comparison of different kinds or sizes a standard method of testing is necessary and the values must be expressed in some defined units. For these reasons laboratory experiments if properly conducted have many advantages over any other method.

One object of such investigation is to find unit values for strength and stiffness, etc. These, because of the complex structure of wood, cannot have a constant value which will be exactly repeated in each test, even though no error be made. The most that can be accomplished is to find average values, the amount of variation above and below, and the laws which govern the variation. On account of the great variability in strength of different specimens of wood even from the same stick and appearing to be alike, it is important to eliminate as far as possible all extraneous factors liable to influence the results of the tests.

The mechanical properties of wood considered in this book are: (1) stiffness and elasticity, (2) tensile strength, (3) compressive or crushing strength, (4) shearing strength, (5) transverse or bending strength, (6) toughness, (7) hardness, (8) cleavability, (9) resilience. In connection with these, associated properties of importance are briefly treated.

In making use of figures indicating the strength or other mechanical properties of wood for the purpose of comparing the relative merits of different species, the fact should be borne in mind that there is a considerable range in variability of each individual material and that small differences, such as a few hundred pounds in values of 10,000 pounds, cannot be considered as a criterion of the quality of the timber. In testing material of the same kind and grade, differences of 25 per cent between individual specimens may be expected in conifers and 50 per cent or even more in hardwoods. The figures given in the tables should be taken as indications rather than fixed values, and as applicable to a large number collectively and not to individual pieces.

Fundamental Considerations and Definitions

Study of the mechanical properties of a material is concerned mostly with its behavior in relation to stresses and strains, and the factors affecting this behavior. A **stress** is a distributed force and may be defined as the mutual action (1) of one body upon another, or (2) of one part of a body upon another part. In the first case the stress is *external;* in the other *internal.* The same stress may be internal from one point of view and external from another. An external force is always balanced by the internal stresses when the body is in equilibrium.

If no external forces act upon a body its particles assume certain relative positions, and it has what is called its *natural shape and size.* If sufficient external force is applied the natural shape and size will be changed. This distortion or deformation of the material is known as the **strain.** Every stress produces a corresponding strain, and within a certain limit (see *elastic limit,* page 5) the strain is directly proportional to the stress producing it.* The same intensity of stress, however, does not

---

* This is in accordance with the discovery made in 1678 by Robert Hooke, and is known as *Hooke's law.*

produce the same strain in different materials or in different qualities of the same material. No strain would be produced in a perfectly rigid body, but such is not known to exist.

Stress is measured in pounds (or other unit of weight or force). A **unit stress** is the stress on a unit of the sectional area. (Unit stress $= \dfrac{P}{A}$) For instance, if a load (P) of one hundred pounds is uniformly supported by a vertical post with a cross-sectional area (A) of ten square inches, the unit compressive stress is ten pounds per square inch.

Strain is measured in inches (or other linear unit). A **unit strain** is the strain per unit of length. Thus if a post 10 inches long before compression is 9.9 inches long under the compressive stress, the total strain is 0.1 inch, and the unit strain is $\dfrac{l}{L} = \dfrac{0.1}{10} = 0.01$ inch per inch of length.

As the stress increases there is a corresponding increase in the strain. This ratio may be graphically shown by means of a diagram or curve plotted with the increments of load or stress as ordinates and the increments of strain as abscissæ. This is known as the **stress-strain diagram.** Within the limit mentioned above the diagram is a straight line. (See Fig. 1.) If the results of similar experiments on different specimens are plotted to the same scales, the diagrams furnish a ready means for comparison. The greater the resistance a material offers to deformation the steeper or nearer the vertical axis will be the line.

There are three kinds of internal stresses, namely, (1) **tensile,** (2) **compressive,** and (3) **shearing.** When external forces act upon a bar in a direction away from its ends or a direct pull, the stress is a tensile stress; when toward the ends or a direct push, compressive stress. In the first instance the strain is an *elongation;* in the second a *shortening.* Whenever the forces tend to cause one portion of the material to slide upon another adjacent to it the action is called a *shear.* The action is that of an ordinary pair of shears. When riveted plates slide on each other the rivets are sheared off.

These three simple stresses may act together, producing compound stresses, as in flexure. When a bow is bent there is a compression of the fibres on the inner or concave side and an elongation of the fibres on the outer or convex side. There is also a tendency of the various fibres to slide past one another in a longitudinal direction. If the bow were made of two or more

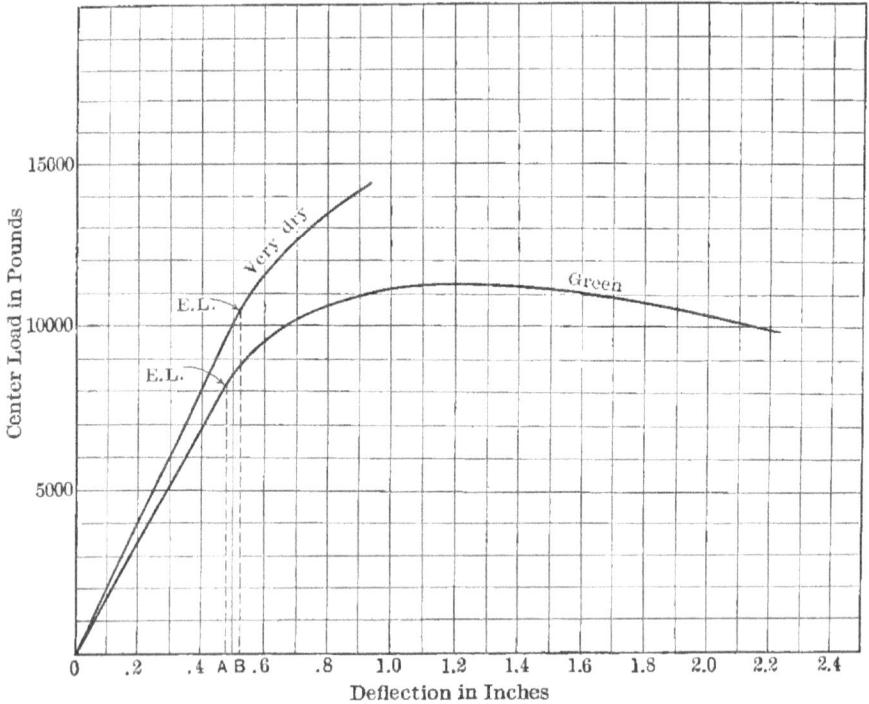

FIG. 1.—Stress-strain diagrams of two longleaf pine beams. E.L. = elastic limit. The areas of the triangles 0(EL)A and 0(EL)B represent the elastic resilience of the dry and green beams, respectively.

separate pieces of equal length it would be noted on bending that slipping occurred along the surfaces of contact, and that the ends would no longer be even. If these pieces were securely glued together they would no longer slip, but the tendency to do so would exist just the same. Moreover, it would be found in the latter case that the bow would be much harder to bend than where the pieces were not glued together—in other words, the *stiffness* of the bow would be materially increased.

Stiffness is the property by means of which a body acted upon by external forces tends to retain its natural size and shape, or resists deformation. Thus a material that is difficult to bend or otherwise deform is stiff; one that is easily bent or otherwise deformed is *flexible*. Flexibility is not the exact counterpart of stiffness, as it also involves toughness and pliability.

If successively larger loads are applied to a body and then removed it will be found that at first the body completely regains its original form upon release from the stress—in other words, the body is **elastic.** No substance known is perfectly elastic, though many are practically so under small loads. Eventually a point will be reached where the recovery of the specimen is incomplete. This point is known as the **elastic limit,** which may be defined as the limit beyond which it is impossible to carry the distortion of a body without producing a permanent alteration in shape. After this limit has been exceeded, the size and shape of the specimen after removal of the load will not be the same as before, and the difference or amount of change is known as the **permanent set.**

Elastic limit as measured in tests and used in design may be defined as that unit stress at which the deformation begins to increase in a faster ratio than the applied load. In practice the elastic limit of a material under test is determined from the stress-strain diagram. It is that point in the line where the diagram begins perceptibly to curve.* (See Fig. 1, page 4.)

Resilience is the amount of work done upon a body in deforming it. Within the elastic limit it is also a measure of the potential energy stored in the material and represents the amount of work the material would do upon being released from a state of stress. This may be graphically represented by a diagram in which the abscissæ represent the amount of deflection and the ordinates the force acting. The area included between the stress-strain curve and the initial line (which is zero) represents the work done. (See Fig. 1, page 4.) If the unit of space is in inches and the unit

---

* If the straight portion does not pass through the origin, a parallel line should be drawn through the origin, and the load at elastic limit taken from this line. (See Fig. 32, page 101.)

of force is in pounds the result is inch-pounds. If the elastic limit is taken as the apex of the triangle the area of the triangle will represent the **elastic resilience** of the specimen. This amount of work can be applied repeatedly and is perhaps the best measure of the toughness of the wood as a working quality, though it is not synonymous with toughness.

Permanent set is due to the **plasticity** of the material. A perfectly plastic substance would have no elasticity and the smallest forces would cause a set. Lead and moist clay are nearly plastic and wood possesses this property to a greater or less extent. The plasticity of wood is increased by wetting, heating, and especially by steaming and boiling. Were it not for this property it would be impossible to dry wood without destroying completely its cohesion, due to the irregularity of shrinkage.

A substance that can undergo little change in shape without breaking or rupturing is **brittle.** Chalk and glass are common examples of brittle materials. Sometimes the word *brash* is used to describe this condition in wood. A brittle wood breaks suddenly with a clean instead of a splintery fracture and without warning. Such woods are unfitted to resist shock or sudden application of load.

The measure of the stiffness of wood is termed the **modulus of elasticity** (or *coefficient of elasticity*). It is the ratio of stress per unit of area to the deformation per unit of length. $\left(E = \dfrac{\text{unit stress}}{\text{unit strain}} \right)$ It is a number indicative of stiffness, not of strength, and only applies to conditions within the elastic limit. It is nearly the same whether derived from compression tests or from tension tests.

A large modulus indicates a stiff material. Thus in green wood tested in static bending it varies from 643,000 pounds per square inch for arborvitæ to 1,662,000 pounds for longleaf pine, and 1,769,000 pounds for pignut hickory. (See Table IX, page 27.) The values derived from tests of small beams of dry material are much greater, approaching 3,000,000 for some of our woods. These values are small when compared with steel which has a modulus of elasticity of about 30,000,000 pounds per square inch. (See Table I.)

TABLE I

COMPARATIVE STRENGTH OF IRON, STEEL, AND WOOD

MATERIAL	Sp. gr., dry	Modulus of elasticity in bending	Tensile strength	Crushing strength	Modulus of rupture
		Lbs. per sq. in.	Lbs. per sq. in.	Lbs. per sq. in.	Lbs. per sq. in.
Cast iron, cold blast (Hodgkinson)......	7.1	17,270,000	16,700	106,000	38,500
Bessemer steel, high grade (Fairbain)....	7.8	29,215,000	88,400	225,600	.....
Longleaf pine, 3.5% moisture (U. S.)....	.63	2,800,000	.....	13,000	21,000
Red spruce, 3.5% moisture (U. S.)........	.41	1,800,000	.....	8,800	14,500
Pignut hickory, 3.5% moisture (U. S.)....	.86	2,370,000	.....	11,130	24,000

NOTE.—Great variation may be found in different samples of metals as well as of wood. The examples given represent reasonable values.

TENSILE STRENGTH

Tension results when a pulling force is applied to opposite ends of a body. This external pull is communicated to the interior, so that any portion of the material exerts a pull or tensile force upon the remainder, the ability to do so depending upon the property of cohesion. The result is an elongation or stretching of the material in the direction of the applied force. The action is the opposite of compression.

Wood exhibits its greatest strength in tension parallel to the grain, and it is very uncommon in practice for a specimen to be pulled in two lengthwise. This is due to the difficulty of making the end fastenings secure enough for the full tensile strength to be brought into play before the fastenings shear off longitudinally. This is not the case with metals, and as a result they are used in almost all places where tensile strength is particularly needed, even though the remainder of the structure, such as sills, beams, joists, posts, and flooring, may be of wood. Thus in a wooden truss bridge the tension members are steel rods.

The tensile strength of wood parallel to the grain depends upon the strength of the fibres and is affected not only by the nature and dimensions of the wood elements but also by their arrangement. It is greatest in straight-grained specimens with thick-walled fibres. Cross grain of any kind materially reduces the tensile strength of wood, since the tensile strength at right angles to the grain is only a small fraction of that parallel to the grain.

TABLE II

RATIO OF STRENGTH OF WOOD IN TENSION AND IN COMPRESSION

(Bul. 10, U. S. Div. of Forestry, p. 44)

KIND OF WOOD	Ratio: $R = \dfrac{\text{Tensile strength}}{\text{compressive strength}}$	A stick 1 square inch in cross section. Weight required to—	
		Pull apart	Crush endwise
Hickory...............	3.7	32,000	8,500
Elm...................	3.8	29,000	7,500
Larch.................	2.3	19,400	8,600
Longleaf pine.........	2.2	17,300	7,400

NOTE.—Moisture condition not given.

Failure of wood in tension parallel to the grain occurs sometimes in flexure, especially with dry material. The tension portion of the fracture is nearly the same as though the piece were pulled in two lengthwise. The fibre walls are torn across obliquely and usually in a spiral direction. There is practically no pulling apart of the fibres, that is, no separation of the fibres along their walls, regardless of their thickness. The nature of tension failure is apparently not affected by the moisture condition of the specimen, at least not so much so as the other strength values.*

Tension at right angles to the grain is closely related to cleavability. When wood fails in this manner the thin fibre walls are torn in two lengthwise while the thick-walled fibres are usually pulled apart along the primary wall.

---

* See Brush, Warren D.: A microscopic study of the mechanical failure of wood. Vol. II, Rev. F. S. Investigations, Washington, D. C., 1912, p. 35.

TABLE III

TENSILE STRENGTH AT RIGHT ANGLES TO THE GRAIN OF SMALL CLEAR PIECES
OF 25 WOODS IN GREEN CONDITION

(Forest Service Cir. 213)

COMMON NAME OF SPECIES	When surface of failure is radial	When surface of failure is tangential
	Lbs. per sq. inch	Lbs. per sq. inch
Hardwoods		
Ash, white	645	671
Basswood	226	303
Beech	633	969
Birch, yellow	446	526
Elm, slippery	765	832
Hackberry	661	786
Locust, honey	1,133	1,445
Maple, sugar	610	864
Oak, post	714	924
red	639	874
swamp white	757	909
white	622	749
yellow	728	929
Sycamore	540	781
Tupelo	472	796
Conifers		
Arborvitæ	241	235
Cypress, bald	242	251
Fir, white	213	304
Hemlock	271	323
Pine, longleaf	240	298
red	179	205
sugar	239	304
western yellow	230	252
white	225	285
Tamarack	236	274

COMPRESSIVE OR CRUSHING STRENGTH

Compression across the grain is very closely related to hardness and transverse shear. There are two ways in which wood is subjected to stress of this kind, namely, (1) with the load acting over the entire area of the specimen, and (2) with a load concentrated over a portion of the area. (See Fig. 2.) The latter is the condition more commonly met with in practice, as, for example, where a post rests on a horizontal sill, or a rail rests on a cross-tie. The former condition, however, gives the true resistance of the grain to simple crushing.

The first effect of compression across the grain is to compact the fibres, the load gradually but irregularly increasing as the density of the material is increased. If the specimen lies on a flat surface and the load is applied to only a portion of the upper area, the bearing plate indents the wood, crushing the upper fibres without affecting the lower part. (See Fig. 3.) As the load increases the projecting ends sometimes split horizontally. (See Fig. 4.) The irregularities in the load are

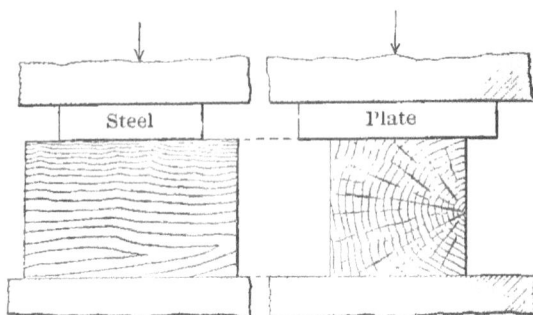

FIG. 2.—Compression across the grain.

FIG. 3.—Side view of failures in compression across the grain, showing crushing of blocks under bearing plate. Specimen at right shows splitting at ends.

due to the fact that the fibres collapse a few at a time, beginning with those with the thinnest walls. | The projection

FIG. 4.—End view of failures in compression across the grain, showing splitting of the ends of the test specimens.

FIG. 5.—Testing a buggy spoke in endwise compression, illustrating the failure by sidewise bending of a long column fixed only at the lower end.

of the ends increases the strength of the material directly beneath the compressing weight by introducing a beam action which helps support the load. This influence is exerted for a short distance only.

When wood is used for columns, props, posts, and spokes, the weight of the load tends to shorten the material endwise. This is **endwise compression,** or compression parallel to the grain. In the case of long columns, that is, pieces in which the length is very great compared with their diameter, the failure is by sidewise bending or flexure, instead of by crushing or splitting. (See Fig. 5.) A familiar instance of this action is afforded by a flexible walking-stick. If downward pressure is exerted with the hand on the upper end of the stick placed vertically on the floor, it will be noted that a definite amount of force must be applied in each instance before decided flexure takes place. After this point is reached a very slight increase of pressure very largely increases the deflection, thus obtaining so great a leverage about the middle section as to cause rupture.

The lateral bending of a column produces a combination of bending with compressive stress over the section, the compressive stress being maximum at the section of greatest deflection on the concave side. The convex surface is under tension, as in an ordinary beam test. (See Fig. 6.) If the same stick is braced in such a way that flexure is prevented, its supporting strength is increased enormously, since the compressive stress acts uniformly over the section, and failure is by crushing or splitting, as in small blocks. In all columns free to bend in any direction the deflection will be seen in the direction in which the column is least stiff. This sidewise bending can be overcome by making pillars and columns thicker in the middle than at the ends, and by bracing studding, props, and compression

F I G. 6.— Unequal distribution of stress in a long column due to lateral bending.

TABLE IV

RESULTS OF COMPRESSION TESTS ACROSS THE GRAIN ON 51 WOODS IN GREEN
CONDITION, AND COMPARISON WITH WHITE OAK

(U. S. Forest Service)

COMMON NAME OF SPECIES	Fibre stress at elastic limit perpendicular to grain	Fibre stress in per cent of white oak, or 853 pounds per sq. in.
	Lbs. per sq. inch	Per cent
Osage orange	2,260	265.0
Honey locust	1,684	197.5
Black locust	1,426	167.2
Post oak	1,148	134.6
Pignut hickory	1,142	133.9
Water hickory	1,088	127.5
Shagbark hickory	1,070	125.5
Mockernut hickory	1,012	118.6
Big shellbark hickory	997	116.9
Bitternut hickory	986	115.7
Nutmeg hickory	938	110.0
Yellow oak	857	100.5
White oak	853	100.0
Bur oak	836	98.0
White ash	828	97.1
Red oak	778	91.2
Sugar maple	742	87.0
Rock elm	696	81.6
Beech	607	71.2
Slippery elm	599	70.2
Redwood	578	67.8
Bald cypress	548	64.3
Red maple	531	62.3
Hackberry	525	61.6
Incense cedar	518	60.8
Hemlock	497	58.3
Longleaf pine	491	57.6
Tamarack	480	56.3
Silver maple	456	53.5
Yellow birch	454	53.2
Tupelo	451	52.9
Black cherry	444	52.1
Sycamore	433	50.8
Douglas fir	427	50.1
Cucumber tree	408	47.8
Shortleaf pine	400	46.9
Red pine	358	42.0
Sugar pine	353	41.4
White elm	351	41.2
Western yellow pine	348	40.8
Lodgepole pine	348	40.8
Red spruce	345	40.5
White pine	314	36.8
Engelman spruce	290	34.0
Arborvitæ	288	33.8
Largetooth aspen	269	31.5
White spruce	262	30.7
Butternut	258	30.3
Buckeye (yellow)	210	24.6
Basswood	209	24.5
Black willow	193	22.6

members of trusses.   The strength of a column also depends
to a considerable extent upon whether the ends are free to
turn or are fixed.

TABLE V

RELATION OF FIBRE STRESS AT ELASTIC LIMIT (r) IN BENDING TO THE CRUSHING
STRENGTH (C) OF BLOCKS CUT THEREFROM, IN POUNDS PER SQUARE INCH

(Forest Service Bul. 70, p. 90)

LONGLEAF PINE

MOISTURE CONDITION	Soaked 50 per cent	Green 23 per cent	14 per cent	11.5 per cent	9.5 per cent	Kiln-dry 6.2 per cent
Number of tests averaged.	5	5	5	5	4	5
r in bending.............	4,920	5,944	6,924	7,852	9,280	11,550
C in compression..	4,668	5,100	6,466	7,466	8,985	10,910
Per cent r is in excess of C.	5.5	16.5	7.1	5.2	3.3	5.9

SPRUCE

MOISTURE CONDITION	Soaked 30 per cent	Green 30 per cent	10 per cent	8.1 per cent	Kiln-dry 3.9 per cent
Number of tests averaged.......	5	4	5	3	4
r in bending.................	3,002	3,362	6,458	8,400	10,170
C in compression.............	2,680	3,025	6,120	7,610	9,335
Per cent r is in excess of C	12.0	11.1	5.5	10.4	9.0

The complexity of the computations depends upon the way in
which the stress is applied and the manner in which the stick
bends.   Ordinarily where the length of the test specimen is not
greater than four diameters and the ends are squarely faced
(see Fig. 7), the force acts uniformly over each square inch of area
and the crushing strength is equal to the maximum load (P)
divided by the area of the cross-section (A). $\left(C = \dfrac{P}{A} \right)$

It has been demonstrated* that the ultimate strength in compression parallel to the grain is very nearly the same as the extreme

---

* See Circular No. 18, U. S. Division of Forestry: Progress in timber
physics, pp. 13–18; also Bulletin 70, U. S. Forest Service: Effect of moisture
on the strength and stiffness of wood, pp. 42, 89–90.

fibre stress at the elastic limit in bending. (See Table V.) In other words, the transverse strength of beams at elastic limit is practically equal to the compressive strength of the same material in short columns. It is accordingly possible to calculate the approximate breaking strength of beams from the compressive strength of short columns except when the wood is brittle. Since tests on endwise compression are simpler, easier to make, and less expensive than transverse bending tests, the importance of this relation is obvious, though it does not do away with the necessity of making beam tests.

When a short column is compressed until it breaks, the manner of failure depends partly upon the anatomical structure and partly upon the degree of humidity of the wood. The fibres (tracheids in conifers) act as hollow tubes bound closely together, and in giving way they either (1) buckle, or (2) bend.*

The first is typical of any dry thin-walled cells, as is usually the case in seasoned white pine and spruce, and in the early wood of hard pines, hemlock, and other species with decided contrast between the two portions of the growth ring. As a rule buckling of a tracheid begins at the bordered pits which form places of least resistance in the walls. In hardwoods such

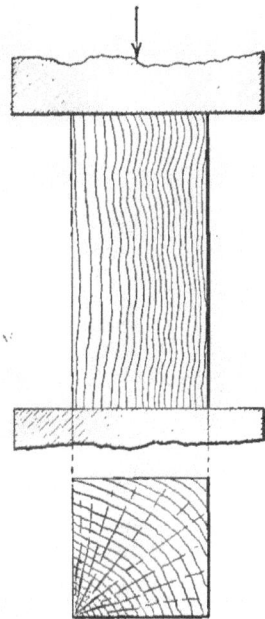

FIG. 7.—Endwise compression of a short column.

as oak, chestnut, ash, etc., buckling occurs only in the thinnest-walled elements, such as the vessels, and not in the true fibres.

According to Jaccard † the folding of the cells is accompanied by characteristic alterations of their walls which seem to split them into extremely thin layers. When greatly magnified, these layers appear in longitudinal sections as delicate threads without

---

* See Bulletin 70, *op. cit.*, p. 129.

† Jaccard, P.: Étude anatomique des bois comprimés. Mit. d. Schw. Centralanstalt f. d. forst. Versuchswesen. X. Band, 1. Heft. Zurich, 1910, p. 66.

TABLE VI

RESULTS OF ENDWISE COMPRESSION TESTS ON SMALL CLEAR PIECES OF 40
WOODS IN GREEN CONDITION

(Forest Service Cir. 213)

Common name of species	Fibre stress at elastic limit	Crushing strength	Modulus of elasticity
	Lbs. per sq. inch	Lbs. per sq. inch	Lbs. per sq. inch
Hardwoods			
Ash, white	3,510	4,220	1,531,000
Basswood	780	1,820	1,016,000
Beech	2,770	3,480	1,412,000
Birch, yellow	2,570	3,400	1,915,000
Elm, slippery	3,410	3,990	1,453,000
Hackberry	2,730	3,310	1,068,000
Hickory, big shellbark	3,570	4,520	1,658,000
bitternut	4,330	4,570	1,616,000
mockernut	3,990	4,320	1,359,000
nutmeg	3,620	3,980	1,411,000
pignut	3,520	4,820	1,980,000
shagbark	3,730	4,600	1,943,000
water	3,240	4,660	1,926,000
Locust, honey	4,300	4,970	1,536,000
Maple, sugar	3,040	3,670	1,463,000
Oak, post	2,780	3,330	1,062,000
red	2,290	3,210	1,295,000
swamp white	3,470	4,360	1,489,000
white	2,400	3,520	946,000
yellow	2,870	3,700	1,465,000
Osage orange	3,980	5,810	1,331,000
Sycamore	2,320	2,790	1,073,000
Tupelo	2,280	3,550	1,280,000
Conifers			
Arborvitæ	1,420	1,990	754,000
Cedar, incense	2,710	3,030	868,000
Cypress, bald	3,560	3,960	1,738,000
Fir, alpine	1,660	2,060	882,000
amabilis	2,763	3,040	1,579,000
Douglas	2,390	2,920	1,440,000
white	2,610	2,800	1,332,000
Hemlock	2,110	2,750	1,054,000
Pine, lodgepole	2,290	2,530	1,219,000
longleaf	3,420	4,280	1.890,000
red	2,470	3,080	1,646,000
sugar	2,340	2,600	1,029,000
western yellow	2,100	2,420	1,271,000
white	2,370	2,720	1,318,000
Redwood	3,420	3,820	1,175,000
Spruce, Engelmann	1,880	2,170	1,021,000
Tamarack	3,010	3,480	1,596,000

any definite arrangements, while on cross section they appear as numerous concentric strata. This may be explained on the ground that the growth of a fibre is by successive layers which, under the influence of compression, are sheared apart. This is particularly the case with thick-walled cells such as are found in late wood.

The second case, where the fibres bend with more or less regular curves instead of buckling, is characteristic of any green or wet wood, and in dry woods where the fibres are thick-walled. In woods in which the fibre walls show all gradations of thickness— in other words, where the transition from the thin-walled cells of

Fig. 8.—Failures of short columns of green spruce.

the early wood to the thick-walled cells of the late wood is gradual —the two kinds of failure, namely, buckling and bending, grade into each other. In woods with very decided contrast between early and late wood the two forms are usually distinct. Except in the case of complete failure the cavity of the deformed cells remains open, and in hardwoods this is true not only of the wood fibres but also of the tube-like vessels. In many cases longitudinal splits occur which isolate bundles of elements by greater or less intervals. The splitting occurs by a tearing of the fibres or rays and not by the separation of the rays from the adjacent elements.

Moisture in wood decreases the stiffness of the fibre walls and enlarges the region of failure. The curve which the fibre walls

make in the region of failure is more gradual and also more irregular than in dry wood, and the fibres are more likely to be separated.

In examining the lines of rupture in compression parallel to the
grain it appears that there does not exist any specific type, that is,
one that is characteristic of all woods. Test blocks taken from
different parts of the same log may show very decided differences

FIG. 9.—Failures of short columns of dry chestnut.

in the manner of failure, while blocks that are much alike in the
size, number, and distribution of the elements of unequal resistance may behave very similarly. The direction of rupture
is, according to Jaccard, not influenced by the distribution of the
medullary rays.* These are curved with the bundles of fibres
to which they are attached. In any case the failure starts at

---

* This does not correspond exactly with the conclusions of A. Thil, who
says ("Constitution anatomique du bois," pp. 140–141): "The sides of the
medullary rays sometimes produce planes of least resistance varying in size
with the height of the rays. The medullary rays assume a direction more
or less parallel to the lumen of the cells on which they border; the latter curve
to the right or left to make room for the ray and then close again beyond it.
If the force acts parallel to the axis of growth, the tracheids are more likely
to be displaced if the marginal cells of the medullary rays are provided with
weak walls that are readily compressed. This explains why on the radial
surface of the test blocks the plane of rupture passes in a direction nearly
following a medullary ray, whereas on the tangential surface the direction
of the plane of rupture is oblique—but with an obliquity varying with the
species and determined by the pitch of the spirals along which the medullary
rays are distributed in the stem." See Jaccard, *op. cit.*, pp. 57 *et seq.*

the weakest points and follows the lines of least resistance. The plane of failure, as visible on radial surfaces, is horizontal, and on the tangential surface it is diagonal.

SHEARING STRENGTH

Whenever forces act upon a body in such a way that one portion tends to slide upon another adjacent to it the action is called a **shear.*** In wood this shearing action may be (1) **along the grain,** or (2) **across the grain.** A tenon breaking out its mortise is a familiar example of shear along the grain, while the shoving off of the tenon itself would be shear across the grain. The use of wood for pins or treenails involves resistance to shear across the grain. Another common instance of the latter is where the steel edge of the eye of an axe or hammer tends to cut off the handle. In Fig. 10 the action of the wooden strut tends to shear off along the grain the portion AB of the wooden tie rod, and it is essential that the length of this portion be great enough to guard against it. Fig. 11 shows characteristic failures in shear along the grain.

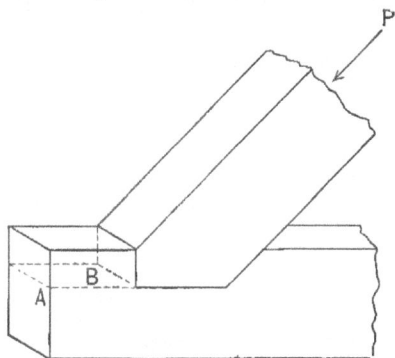

FIG. 10.—Example of shear along the grain.

FIG. 11.—Failures of test specimens in shear along the grain. In the block at the left the surface of failure is radial; in the one at the right, tangenital.

* Shear should not be confused with ordinary cutting or incision.

TABLE VIl

SHEARING STRENGTH ALONG THE GRAIN OF SMALL CLEAR PIECES OF 41 WOODS
IN GREEN CONDITION

(Forest Service Cir. 213)

COMMON NAME OF SPECIES	When surface of failure is radial	When surface of failure is tangential
	Lbs. per sq. inch	Lbs. per sq. inch
Hardwoods		
Ash, black	876	832
white	1,360	1,312
Basswood	560	617
Beech	1,154	1,375
Birch, yellow	1,103	1,188
Elm, slippery	1,197	1,174
white	778	872
Hackberry	1,095	1,161
Hickory, big shellbark	1,134	1,191
bitternut	1,134	1,348
mockernut	1,251	1,313
nutmeg	1,010	1,053
pignut	1,334	1,457
shagbark	1,230	1,297
water	1,390	1,490
Locust, honey	1,885	2,096
Maple, red	1,130	1,330
sugar	1,193	1,455
Oak, post	1,196	1,402
red	1,132	1,195
swamp white	1,198	1,394
white	1,096	1,292
yellow	1,162	1,196
Sycamore	900	1,102
Tupelo	978	1,084
Conifers		
Arborvitæ	617	614
Cedar, incense	613	662
Cypress, bald	836	800
Fir, alpine	573	654
amabilis	517	639
Douglas	853	858
white	742	723
Hemlock	790	813
Pine, lodgepole	672	747
longleaf	1,060	953
red	812	741
sugar	702	714
western yellow	686	706
white	649	639
Spruce, Engelmann	607	624
Tamarack	883	843

Both shearing stresses may act at the same time. Thus the weight carried by a beam tends to shear it off at right angles to the axis; this stress is equal to the resultant force acting perpendicularly at any point, and in a beam uniformly loaded and

FIG. 12.—Horizontal shear in a beam.

supported at either end is maximum at the points of support and zero at the centre. In addition there is a shearing force tending to move the fibres of the beam past each other in a longitudinal direction. (See Fig. 12.) This longitudinal shear is maximum at the neutral plane and decreases toward the upper and lower surfaces.

Shearing across the grain is so closely related to compression at right angles to the grain and to hardness that there is little to be gained by making separate tests upon it. Knowledge of shear parallel to the grain is important, since wood frequently fails in that way. The value of shearing stress parallel to the grain is found by dividing the maximum load in pounds (P) by the area of the cross section in inches (A).

F I G. 13.—Oblique shear in a short column.

FIG. 14.—Failure of short column by oblique shear.

$$\left(\text{Shear} = \frac{P}{A}\right)$$

Oblique shearing stresses are developed in a bar when it is subjected to direct tension or compression. The maximum shearing stress occurs along a plane when it makes an angle of 45 degrees with the axis of the specimen. In this case, shear = $\frac{P}{2A}$. When the value of the angle θ is less than 45 degrees, the shear along the plane = $\frac{P}{A}$ sin θ cos θ. (See Fig. 13.) The effect of oblique shear is often visible in the failures of short columns. (See Fig. 14.)

TABLE VIII

SHEARING STRENGTH ACROSS THE GRAIN OF VARIOUS AMERICAN WOODS

(J. C. Trautwine. Jour. Franklin Institute. Vol. 109, 1880, pp. 105 106)

KIND OF WOOD	Lbs. per sq. inch	KIND OF WOOD	Lbs. per sq. inch
Ash	6,280	Hickory	7,285
Beech	5,223	Locust	7,176
Birch	5,595	Maple	6,355
Cedar (white)	1,372	Oak	4,425
Cedar (white)	1,519	Oak (live)	8,480
Cedar (Central Amer.)	3,410	Pine (white)	2,480
Cherry	2,945	Pine (northern yellow)	4,340
Chestnut	1,536	Pine (southern yellow)	5,735
Dogwood	6,510	Pine (very resinous yellow)	5,053
Ebony	7,750	Poplar	4,418
Gum	5,890	Spruce	3,255
Hemlock	2,750	Walnut (black)	4,728
Hickory	6,045	Walnut (common)	2,830

Note.—Two specimens of each were tested. All were fairly seasoned and without defects. The piece sheared off was ⅝ in. The single circular area of each pin was 0.322 sq. in.

TRANSVERSE OR BENDING STRENGTH: BEAMS

When external forces acting in the same plane are applied at right angles to the axis of a bar so as to cause it to bend, they occasion a shortening of the longitudinal fibres on the concave side and an elongation of those on the convex side. Within the elastic limit the relative stretching and contraction of the fibres is directly * proportional to their distances from a plane inter-

---

* While in reality this relationship does not exactly hold, the formulæ for beams are based on its assumption.

mediate between them—the **neutral plane.** (N_1P in Fig. 15.)
Thus the fibres half-way between the neutral plane and the
outer surface experience only half as much shortening or elongation as the outermost or extreme fibres. Similarly for other
distances. The elements along the neutral plane experience no
tension or compression in an axial direction. The line of inter-
section of this plane and the plane of section is known as the
neutral axis (NA in Fig. 15) of the section.

If the bar is symmetrical and homogeneous the neutral plane
is located half-way between the upper and lower surfaces, so long
as the deflection does not exceed the elastic limit of the material.

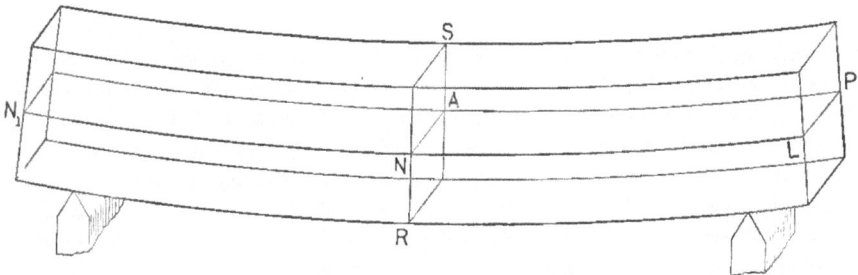

FIG. 15.—Diagram of a simple beam. N_1P = neutral plane, NA = neutral axis of
section RS.

Owing to the fact that the tensile strength of wood is from two
to nearly four times the compressive strength, it follows that at
rupture the neutral plane is much nearer the convex than the
concave side of the bar or beam, since the sum of all the compressive stresses on the concave portion must always equal the
sum of the tensile stresses on the convex portion. The neutral
plane begins to change from its central position as soon as the
elastic limit has been passed. Its location at any time is very
uncertain.

The external forces acting to bend the bar also tend to rupture
it at right angles to the neutral plane by causing one transverse
section to slip past another. This stress at any point is equal
to the resultant perpendicular to the axis of the forces acting
at this point, and is termed the **transverse shear** (or in the case
of beams, **vertical shear**).

In addition to this there is a shearing stress, tending to move the fibres past one another in an axial direction, which is called **longitudinal shear** (or in the case of beams, **horizontal shear**). This stress must be taken into consideration in the design of timber structures. It is maximum at the neutral plane and decreases to zero at the outer elements of the section. The shorter the span of a beam in proportion to its height, the greater is the liability of failure in horizontal shear before the ultimate strength of the beam is reached.

Beams

There are three common forms of beams, as follows:

(1) **Simple beam**—a bar resting upon two supports, one near each end. (See Fig. 16, No. 1.)

(2) **Cantilever beam**—a bar resting upon one support or

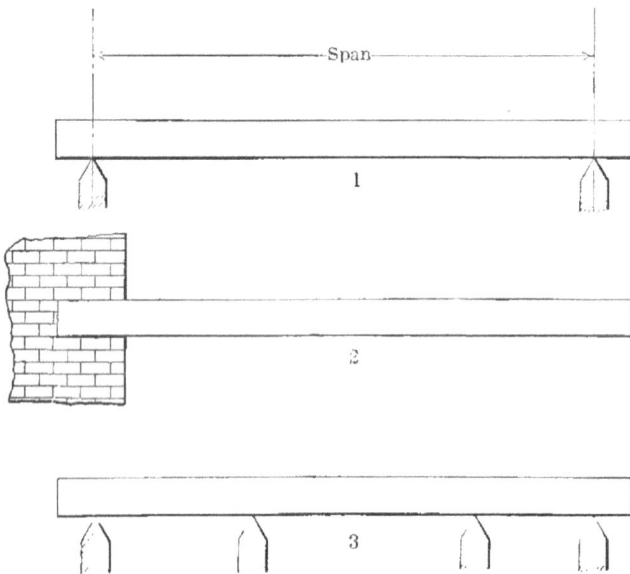

FIG. 16.—Three common forms of beams. 1. Simple. 2. Cantilever. 3. Continuous.

fulcrum, or that portion of any beam projecting out of a wall or beyond a support. (See Fig. 16, No. 2.)

(3) **Continuous beam**—a bar resting upon more than two supports. (See Fig. 16, No. 3.)

Stiffness of Beams

The two main requirements of a beam are stiffness and strength. The formula for the *modulus of elasticity* (*E*) or measure of stiffness of a rectangular prismatic simple beam loaded at the centre and resting freely on supports at either end is:*

$$E = \frac{P' \, l^3}{4 \, D \, b \, h^3}$$

b = breadth or width of beam, inches.
h = height or depth of beam, inches.
l = span (length between points of supports) of beam, inches.
D = deflection produced by load P', inches.
P' = load at or below elastic limit, pounds.

From this formula it is evident that for rectangular beams of the same material, mode of support, and loading, the deflection is affected as follows:

(1) It is inversely proportional to the width for beams of the same length and depth. If the width is tripled the deflection is one-third as great.

(2) It is inversely proportional to the cube of the depth for beams of the same length and breadth. If the depth is tripled the deflection is one twenty-seventh as great.

(3) It is directly proportional to the cube of the span for beams of the same breadth and depth. Tripling the span gives twenty-seven times the deflection.

The number of pounds which concentrated at the centre will deflect a rectangular prismatic simple beam one inch may be found from the preceding formula by substituting $D = 1''$ and solving for P'. The formula then becomes:

$$\text{Necessary weight } (P') = \frac{4 \, E \, b \, h^3}{l^3}$$

In this case the values for E are read from tables prepared from data obtained by experimentation on the given material.

---

* Only this form of beam is considered since it is the simplest. For cantilever and continuous beams, and beams rigidly fixed at one or both ends, as well as for different methods of loading, different forms of cross section, etc., other formulæ are required. See any book on mechanics.

Strength of Beams

The measure of the breaking strength of a beam is expressed in terms of unit stress by a *modulus of rupture*, which is a purely hypothetical expression for points beyond the elastic limit. The formula used in computing this modulus is as follows:

$$R = \frac{1.5\,P\,l}{b\,h^2}$$

b, h, l = breadth, height, and span, respectively, as in preceding formula.

R = modulus of rupture, pounds per square inch.

P = maximum load, pounds.

In calculating the fibre stress at the elastic limit the same formula is used except that the load at elastic limit (P_1) is substituted for the maximum load (P).

From this formula it is evident that for rectangular prismatic beams of the same material, mode of support, and loading, the load which a given beam can support varies as follows:

(1) It is directly proportional to the breadth for beams of the same length and depth, as is the case with stiffness.

(2) It is directly proportional to the square of the height for beams of the same length and breadth, instead of as the cube of this dimension as in stiffness.

(3) It is inversely proportional to the span for beams of the same breadth and depth and not to the cube of this dimension as in stiffness.

The fact that the strength varies as the *square* of the height and the stiffness as the *cube* explains the relationship of bending to thickness. Were the law the same for strength and stiffness a thin piece of material such as a sheet of paper could not be bent any further without breaking than a thick piece, say an inch board.

Kinds of Loads

There are various ways in which beams are loaded, of which the following are the most important:

(1) **Uniform load** occurs where the load is spread evenly over the beam.

TABLE IX

RESULTS OF STATIC BENDING TESTS ON SMALL CLEAR BEAMS OF 49 WOODS
IN GREEN CONDITION

(Forest Service Cir. 213)

COMMON NAME OF SPECIES	Fibre stress at elastic limit	Modulus of rupture	Modulus of elasticity	Work in bending		
				To elastic limit	To maximum load	Total
Hardwoods	Lbs. per sq. in.	Lbs. per sq. in.	Lbs. per sq. in.	In.-lbs. per cu. in.	In.-lbs. per cu. in.	In.-lbs. per cu. in.
Ash, black.........	2,580	6,000	960,000	0.41	13.1	38.9
white........	5,180	9,920	1,416,000	1.10	20.0	43.7
Basswood.........	2,480	4,450	842,000	.45	5.8	8.9
Beech.............	4,490	8,610	1,353,000	.96	14.1	31.4
Birch, yellow......	4,190	8,390	1,597,000	.62	14.2	31.5
Elm, rock.........	4,290	9,430	1,222,000	.90	19.4	47.4
slippery......	5,560	9,510	1,314,000	1.32	11.7	44.2
white........	2,850	6,940	1,052,000	.44	11.8	27.4
Gum, red.........	3,460	6,450	1,138,000	....	....	....
Hackberry.........	3,320	7,800	1,170,000	.56	19.6	52.9
Hickory,						
big shellbark.....	6,370	11,110	1,562,000	1.47	24.3	78.0
bitternut........	5,470	10,280	1,399,000	1.22	20.0	75.5
mockernut.......	6,550	11,110	1,508,000	1.50	31.7	84.4
nutmeg.........	4,860	9,060	1,289,000	1.06	22.8	58.2
pignut..........	5,860	11,810	1,769,000	1.12	30.6	86.7
shagbark........	6,120	11,000	1,752,000	1.22	18.3	72.3
water..........	5,980	10,740	1,563,000	1.29	18.8	52.9
Locust, honey......	6,020	12,360	1,732,000	1.28	17.3	64.4
Maple, red........	4,450	8,310	1,445,000	.78	9.8	17.1
sugar......	4,630	8,860	1,462,000	.88	12.7	32.0
Oak, post.........	4,720	7,380	913,000	1.39	9.1	17.4
red..........	3,490	7,780	1,268,000	.60	11.4	26.0
swamp white..	5,380	9,860	1,593,000	1.05	14.5	37.6
tanbark......	6,580	10,710	1,678,000	1.49	....	....
white........	4,320	8,090	1,137,000	.95	12.1	36.7
yellow.......	5,060	8,570	1,219,000	1.20	11.7	30.7
Osage orange......	7,760	13,660	1,329,000	2.53	37.9	101.7
Sycamore.........	2,820	6,300	964,000	.51	7.1	13.6
Tupelo............	4,300	7,380	1,045,000	1.00	7.8	20.9
Conifers						
Arborvitæ.........	2,600	4,250	643,000	.60	5.7	9.5
Cedar, incense.....	3,950	6,040	754,000	....	....	....
Cypress, bald......	4,430	7,110	1,378,000	.96	5.1	15.4
Fir, alpine........	2,366	4,450	861,000	.66	4.4	7.4
amabilis.......	4,060	6,570	1,323,000	....	....	....
Douglas.......	3,570	6,340	1,242,000	.59	6.6	13.6
white.......	3,880	5,970	1,131,000	.77	5.2	14.9
Hemlock..........	3,410	5,770	917,000	.73	6.6	12.9
Pine, lodgepole.....	3,080	5,130	1,015,000	.54	5.1	7.4
longleaf.......	5,090	8,630	1,662,000	.88	8.1	34.8
red..........	3,740	6,430	1,384,000	.59	5.8	28.0
shortleaf......	4,360	7,710	1,395,000	....	....	....
sugar.........	3,330	5,270	966,000	.66	5.0	11.6
west, yellow...	3,180	5,180	1,111,000	.52	4.3	15.6
White.............	3,410	5,310	1,073,000	.62	5.9	13.3
Redwood..........	4,530	6,560	1,024,000	....	....	....
Spruce, Engelmann.	2,740	4,550	866,000	.50	4.8	6.1
red........	3,440	5,820	1,143,000	.62	6.0	....
white......	3,160	5,200	968,000	.58	6.6	....
Tamarack.........	4,200	7,170	1,236,000	.84	7.2	30.0

(2) **Concentrated load** occurs where the load is applied at single point or points.

(3) **Live** or **immediate load** is one of momentary or short duration at any one point, such as occurs in crossing a bridge.

(4) **Dead** or **permanent load** is one of constant and indeterminate duration, as books on a shelf. In the case of a bridge the weight of the structure itself is the dead load. All large beams support a uniform dead load consisting of their own weight.

The effect of dead load on a wooden beam may be two or more times that produced by an immediate load of the same weight. Loads greater than the elastic limit are unsafe and will generally result in rupture if continued long enough. A beam may be considered safe under permanent load when the deflections diminish during equal successive periods of time. A continual increase in deflection indicates an unsafe load which is almost certain to rupture the beam eventually.

Variations in the humidity of the surrounding air influence the deflection of dry wood under dead load, and increased deflections during damp weather are cumulative and not recovered by subsequent drying. In the case of longleaf pine, dry beams may with safety be loaded permanently to within three-fourths of their elastic limit as determined from ordinary static tests. Increased moisture content, due to greater humidity of the air, lowers the elastic limit of wood so that what was a safe load for the dry material may become unsafe.

When a dead load not great enough to rupture a beam has been removed, the beam tends gradually to recover its former shape, but the recovery is not always complete. If specimens from such a beam are tested in the ordinary testing machine it will be found that the application of the dead load did not affect the stiffness, ultimate strength, or elastic limit of the material. In other words, the deflections and recoveries produced by live loads are the same as would have been produced had not the beam previously been subjected to a dead load.*

* See Tiemann, Harry D.: Some results of dead load bending tests of timber by means of a recording deflectometer. Proc. Am. Soc. for Testing Materials. Phila. Vol. IX, 1909, pp. 534–548.

Maximum load is the greatest load a material will support and is usually greater than the load at rupture.

Safe load is the load considered safe for a material to support in actual practice.   It is always less than the load at elastic limit and is usually taken as a certain proportion of the ultimate or breaking load.

The ratio of the breaking to the safe load is called the factor of safety. $\left(\text{Factor of safety} = \dfrac{\text{ultimate strength}}{\text{safe load}}\right)$ In order to make due allowance for the natural variations and imperfections in wood and in the aggregate structure, as well as for variations in the load, the factor of safety is usually as high as 6 or 10, especially if the safety of human life depends upon the structure. This means that only from one-sixth to one-tenth of the computed strength values is considered safe to use.   If the depth of timbers exceeds four times their thickness there is a great tendency for the material to twist when loaded.   It is to overcome this tendency that floor joists are braced at frequent intervals. Short deep pieces shear out or split before their strength in bending can fully come into play.

Application of Loads

There are three* general methods in which loads may be applied to beams, namely:

(1) **Static loading** or the gradual imposition of load so that the moving parts acquire no appreciable momentum.   Loads are so applied in the ordinary testing machine.

(2) **Sudden imposition of load without initial velocity.** " Thus in the case of placing a load on a beam, if the load be brought into contact with the beam, but its weight sustained by external means, as by a cord, and then this external support be *suddenly* (instantaneously) removed, as by quickly cutting the cord, then, although the load is already touching the beam (and hence there is no real impact), yet the beam is at first offering no resistance, as it has yet suffered no deformation.   Furthermore, as the beam deflects the

---

* A fourth might be added, namely, **vibratory, or harmonic repetition,** which is frequently serious in the case of bridges.

resistance increases, but does not come to be equal to the load until it has attained its normal deflection. In the meantime there has been an unbalanced force of gravity acting, of a constantly diminishing amount, equal at first to the entire load, at the normal deflection. But at this instant the load and the beam are in motion, the hitherto unbalanced force having produced an accelerated velocity, and this velocity of the weight and beam gives to them an energy, or *vis viva*, which must now spend itself in overcoming an *excess* of resistance over and above the imposed load, and the whole mass will not stop until the deflection (as well as the resistance) has come to be equal to *twice* that corresponding to the static load imposed. Hence we say the effect of a suddenly imposed load is to produce twice the deflection and stress of the same load statically applied. It must be evident, however, that this case has nothing in common with either the ordinary ' static ' tests of structural materials in testing-machines, or with impact tests." *

(3) **Impact, shock, or blow.** † There are various common uses of wood where the material is subjected to sudden shocks and jars or impact. Such is the action on the felloes and spokes of a wagon wheel passing over a rough road; on a hammer handle when a blow is struck; on a maul when it strikes a wedge.

Resistance to impact is resistance to energy which is measured by the product of the force into the space through which it moves, or by the product of one-half the moving mass which causes the shock into the square of its velocity. The work done upon the piece at the instant the velocity is entirely removed from the striking body is equal to the total energy of that body. It is impossible, however, to get all of the energy of the striking body stored in the specimen, though the greater the mass and the shorter the space through which it moves, or, in other words, the greater the proportion of weight and the smaller the proportion of velocity making up the energy of the striking body, the more energy the specimen will absorb. The rest is lost in friction, vibrations, heat, and motion of the anvil.

---

* Johnson, J. B.: The materials of construction, pp. 81–82.

† See Tiemann, Harry D.: The theory of impact and its application to testing materials. Jour. Franklin Inst., Oct., Nov., 1909, pp. 235–259, 336–364.

In impact the stresses produced become very complex and difficult to measure, especially if the velocity is high, or the mass of the beam itself is large compared to that of the weight.

The difficulties attending the measurement of the stresses beyond the elastic limit are so great that commonly they are not reckoned. Within the elastic limit the formulæ for calculating the stresses are based on the assumption that the deflection is proportional to the stress in this case as in static tests.

A common method of making tests upon the resistance of wood to shock is to support a small beam at the ends and drop a heavy weight upon it in the middle. (See Fig. 40, page 111.) The height of the weight is increased after each drop and records of the deflection taken until failure. The total work done upon the specimen is equal to the area of the stress-strain diagram plus the effect of local inertia of the molecules at point of contact.

The stresses involved in impact are complicated by the fact that there are various ways in which the energy of the striking body may be spent:

(a) It produces a local deformation of both bodies at the surface of contact, within or beyond the elastic limit. In testing wood the compression of the substance of the steel striking-weight may be neglected, since the steel is very hard in comparison with the wood. In addition to the compression of the fibres at the surface of contact resistance is also offered by the inertia of the particles there, the combined effect of which is a stress at the surface of contact often entirely out of proportion to the compression which would result from the action of a static force of the same magnitude. It frequently exceeds the crushing strength at the extreme surface of contact, as in the case of the swaging action of a hammer on the head of an iron spike, or of a locomotive wheel on the steel rail. This is also the case when a bullet is shot through a board or a pane of glass without breaking it as a whole.

(b) It may move the struck body as a whole with an accelerated velocity, the resistance consisting of the inertia of the body. This effect is seen when a croquet ball is struck with a mallet.

(c) It may deform a fixed body against its external supports and resistances. In making impact tests in the laboratory the test specimen is in reality in the nature of a cushion between two impacting bodies, namely, the striking weight and the base of the machine. It is important that the mass of this base be sufficiently great that its relative velocity to that of the common centre of gravity of itself and the striking weight may be disregarded.

(d) It may deform the struck body as a whole against the resisting stresses developed by its own inertia, as, for example, when a baseball bat is broken by striking the ball.

TABLE X

RESULTS OF IMPACT BENDING TESTS ON SMALL CLEAR BEAMS OF 34 WOODS IN GREEN CONDITION

(Forest Service Cir. 213)

COMMON NAME OF SPECIES	Fibre stress at elastic limit	Modulus of elasticity	Work in bending to elastic limit
	Lbs. per sq. in.	Lbs. per sq. in.	In.-lbs. per cu. in.
Hardwoods			
Ash, black	7,840	955,000	3.69
white	11,710	1,564,000	4.93
Basswood	5,480	917,000	1.84
Beech	11,760	1,501,000	5.10
Birch, yellow	11,080	1,812,000	3.79
Elm, rock	12,090	1,367,000	6.52
slippery	11,700	1,569,000	4.86
white	9,910	1,138,000	4.82
Hackberry	10,420	1,398,000	4.48
Locust, honey	13,460	2,114,000	4.76
Maple, red	11,670	1,411,000	5.45
sugar	11,680	1,680,000	4.55
Oak, post	11,260	1,596,000	4.41
red	10,580	1,506,000	4.16
swamp white	13,280	2,048,000	4.79
white	9,860	1,414,000	3.84
yellow	10,840	1,479,000	4.44
Osage orange	15,520	1,498,000	8.92
Sycamore	8,180	1,165,000	3.22
Tupelo	7,650	1,310,000	2.49
Conifers			
Arborvitæ	5,290	778,000	2.04
Cypress, bald	8,290	1,431,000	2.71
Fir, alpine	5,280	982,000	1.59
Douglas	8,870	1,579,000	2.79
white	7,230	1,326,000	2.21
Hemlock	6,330	1,025,000	2.19
Pine, lodgepole	6,870	1,142,000	2.31
longleaf	9,680	1,739,000	3.02
red	7,480	1,438,000	2.18
sugar	6,740	1,083,000	2.34
western yellow	7,070	1,115,000	2.51
white	6,490	1,156,000	2.06
Spruce, Engelmann	6,300	1,076,000	2.09
Tamarack	7,750	1,263,000	2.67

Impact testing is difficult to conduct satisfactorily and the data obtained are of chief value in a relative sense, that is, for comparing the shock-resisting ability of woods of which like specimens have been subjected to exactly identical treatment. Yet this test is one of the most important made on wood, as it brings out properties not evident from other tests. Defects and brittleness are revealed by impact better than by any other kind of test. In common practice nearly all external stresses are of the nature of impact. In fact, no two moving bodies can come together without impact stress. Impact is therefore the commonest form of applied stress, although the most difficult to measure.

Failures in Timber Beams

If a beam is loaded too heavily it will break or fail in some characteristic manner. These failures may be classified according to the way in which they develop, as tension, compression, and horizontal shear; and according to the appearance of the broken surface, as brash, and fibrous. A number of forms may develop if the beam is completely ruptured.

Since the tensile strength of wood is on the average about three times as great as the compressive strength, a beam should, therefore, be expected to fail by the formation in the first place of a fold on the compression side due to the crushing action, followed by failure on the tension side. This is usually the case in green or moist wood. In dry material the first visible failure is not infrequently on the lower or tension side, and various attempts have been made to explain why such is the case.*

Within the elastic limit the elongations and shortenings are equal, and the neutral plane lies in the middle of the beam. (See page 23.) Later the top layer of fibres on the upper or compression side fail, and on the load increasing, the next layer of fibres fail, and so on, even though this failure may not be visible. As a result the shortenings on the upper side of the beam become considerably greater than the elongations on the lower side. The neutral plane must be presumed to sink gradually toward the tension side, and when the stresses on the outer fibres at the bottom

---

*See Proc. Int. Assn. for Testing Materials, 1912, XXIII₂, pp. 12–13.

have become sufficiently great, the fibres are pulled in two, the tension area being much smaller than the compression area. The rupture is often irregular, as in direct tension tests. Failure may occur partially in single bundles of fibres some time before the final failure takes place. One reason why the failure of a dry beam is different from one that is moist, is that drying increases the stiffness of the fibres so that they offer more resistance to crushing, while it has much less effect upon the tensile strength.

There is considerable variation in tension failures depending upon the toughness or the brittleness of the wood, the arrangement of the grain, defects, etc., making further classification desirable. The four most common forms are:

(1) **Simple tension,** in which there is a direct pulling in two of the wood on the under side of the beam due to a tensile stress parallel to the grain. (See Fig. 17, No. 1.) This is common in straight-grained beams, particularly when the wood is seasoned.

(2) **Cross-grained tension,** in which the fracture is caused by a tensile force acting oblique to the grain. (See Fig. 17, No. 2.) This is a common form of failure where the beam has diagonal, spiral or other form of cross grain on its lower side. Since the tensile strength of wood across the grain is only a small fraction of that with the grain it is easy to see why a cross-grained timber would fail in this manner.

(3) **Splintering tension,** in which the failure consists of a considerable number of slight tension failures, producing a ragged or splintery break on the under surface of the beam. (See Fig. 17, No. 3.) This is common in tough woods. In this case the surface of fracture is fibrous.

(4) **Brittle tension,** in which the beam fails by a clean break extending entirely through it. (See Fig. 17, No. 4.) It is characteristic of a brittle wood which gives way suddenly without warning, like a piece of chalk. In this case the surface of fracture is described as brash.

Compression failure (see Fig. 17, No. 5) has few variations except that it appears at various distances from the neutral plane of the beam. It is very common in green timbers. The compressive stress parallel to the fibres causes them to buckle or bend as in an endwise compressive test. This action usually

begins on the top side shortly after the elastic limit is reached
and extends downward, sometimes almost reaching the neutral
plane before complete failure occurs.   Frequently two or more
failures develop at about the same time.

Horizontal shear failure, in which the upper and lower portions
of the beam slide along each other for a portion of their length

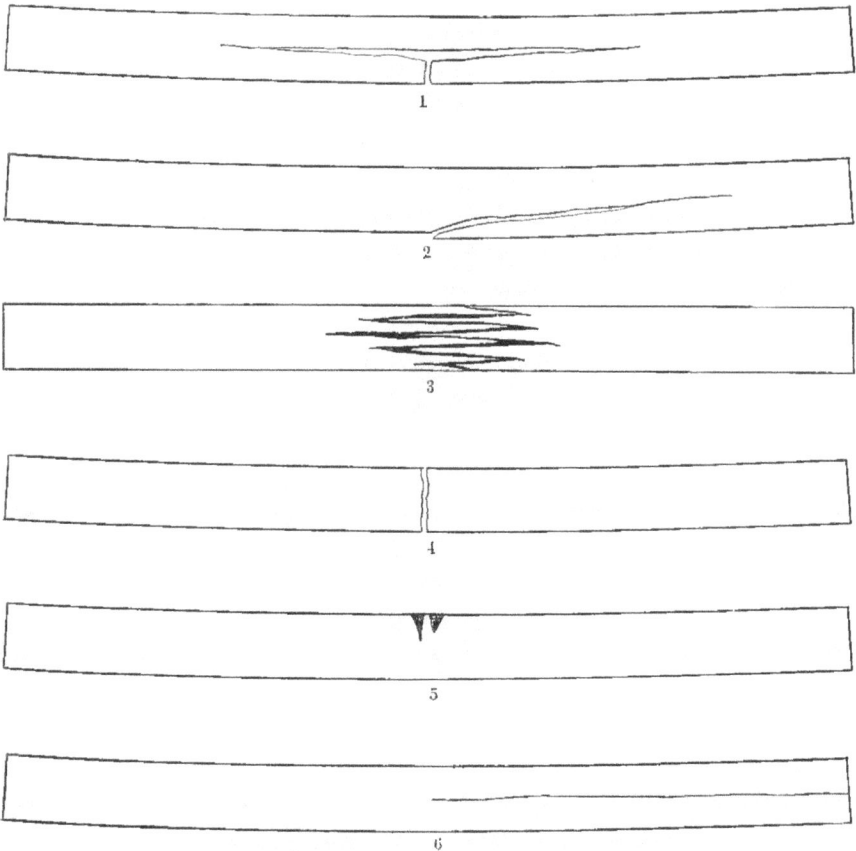

FIG. 17.—Characteristic failures of simple beams.

either at one or at both ends (see Fig. 17, No. 6), is fairly
common in air-dry material and in green material when the ratio
of the height of the beam to the span is relatively large.   It is not
common in small clear specimens.   It is often due to shake or
season checks, common in large timbers, which reduce the actual

TABLE XI

MANNER OF FIRST FAILURE OF LARGE BEAMS

(Forest Service Bul. 108, p. 56)

COMMON NAME OF SPECIES	Total number of tests	Per cent of total failing by		
		Tension	Compression	Shear
Longleaf pine:				
green..........	17	18	24	58
dry............	9	22	22	56
Douglas fir:				
green..........	191	27	72	1
dry............	91	19	76	5
Shortleaf pine:				
green..........	48	27	56	17
dry............	13	54	..	46
Western larch:				
green..........	62	23	71	6
dry............	52	54	19	27
Loblolly pine:				
green..........	111	40	53	7
dry............	25	60	12	28
Tamarack:				
green..........	30	37	53	10
dry............	9	45	22	33
Western hemlock:				
green..........	39	21	74	5
dry............	44	11	66	23
Redwood:				
green..........	28	43	50	7
dry............	12	83	17	..
Norway pine:				
green..........	49	18	76	6
dry............	10	30	60	10

NOTE.—These tests were made on timbers ranging in cross section from 4″ x 10″ to 8″ x 16″, and with a span of 15 feet.

area resisting the shearing action considerably below the calculated area used in the formula for horizontal shear. (See page 98 for this formula.) For this reason it is unsafe, in designing large timber beams, to use shearing stresses higher than those calculated for beams that failed in horizontal shear. The effect of a failure in horizontal shear is to divide the beam into two or more beams the combined strength of which is much less than that of the original beam. Fig. 18 shows a large beam in which two failures in horizontal shear occurred at the same end.   That the parts behave independently is shown by the compression failure below the original location of the neutral plane.

Table XI gives an analysis of the causes of first failure in 840 large timber beams of nine different species of conifers.   Of

Photo by U. S. Forest Service.

FIG. 18.—Failure of a large beam by horizontal shear.

the total number tested 165 were air-seasoned, the remainder green.   The failure occurring first signifies the point of greatest weakness in the specimen under the particular conditions of loading employed (in this case, third-point static loading).

TOUGHNESS: TORSION

Toughness is a term applied to more than one property of wood. Thus wood that is difficult to split is said to be tough.   Again, a tough wood is one that will not rupture until it has deformed considerably under loads at or near its maximum strength, or one which still hangs together after it has been ruptured and may be bent back and forth without breaking apart.   Toughness includes flexibility and is the reverse of brittleness, in that tough woods break gradually and give warning of failure. Tough woods offer great resistance to impact and will permit rougher treatment in manipulations attending manufacture and use. Toughness is dependent upon the strength, cohesion, quality, length, and arrangement of fibre, and the pliability of the wood. Coniferous woods as a rule are not as tough as hardwoods, of which hickory and elm are the best examples.

The torsion or twisting test is useful in determining the toughness of wood. If the ends of a shaft are turned in opposite directions, or one end is turned and the other is fixed, all of the fibres except those at the axis tend to assume the form of helices. (See Fig. 19.) The strain produced by torsion or twisting is essentially

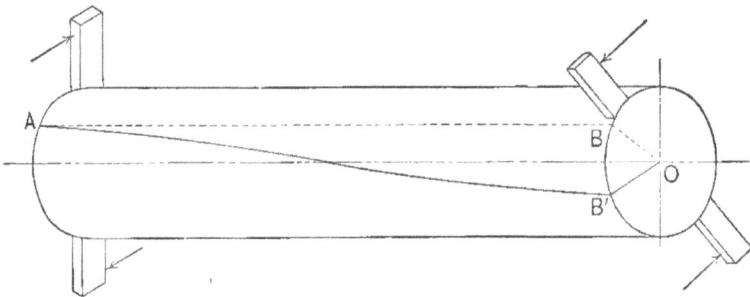

Fig. 19.—Torsion of a shaft.

shear transverse and parallel to the fibres, combined with longitudinal tension and transverse compression. Within the elastic limit the strains increase directly as the distance from the axis of the specimen. The outer elements are subjected to tensile stresses, and as they become twisted tend to compress those near the axis. The elongated elements also contract laterally. Cross sections which were originally plane become warped. With increasing strain the lateral adhesion of the outer fibres is destroyed, allowing them to slide past each other, and reducing greatly their power of resistance. In this way the strains on the fibres nearer the axis are progressively increased until finally all of the elements are sheared apart. It is only in the toughest materials that the full effect of this action can be observed. (See Fig. 20.) Brittle woods snap off suddenly with only a small amount of torsion, and their fracture is irregular and oblique to the axis of the piece instead of frayed out and more nearly perpendicular to the axis as is the case with tough woods.

HARDNESS

The term *hardness* is used in two senses, namely: (1) resistance to indentation, and (2) resistance to abrasion or scratching. In the latter sense hardness combined with toughness is a measure of the

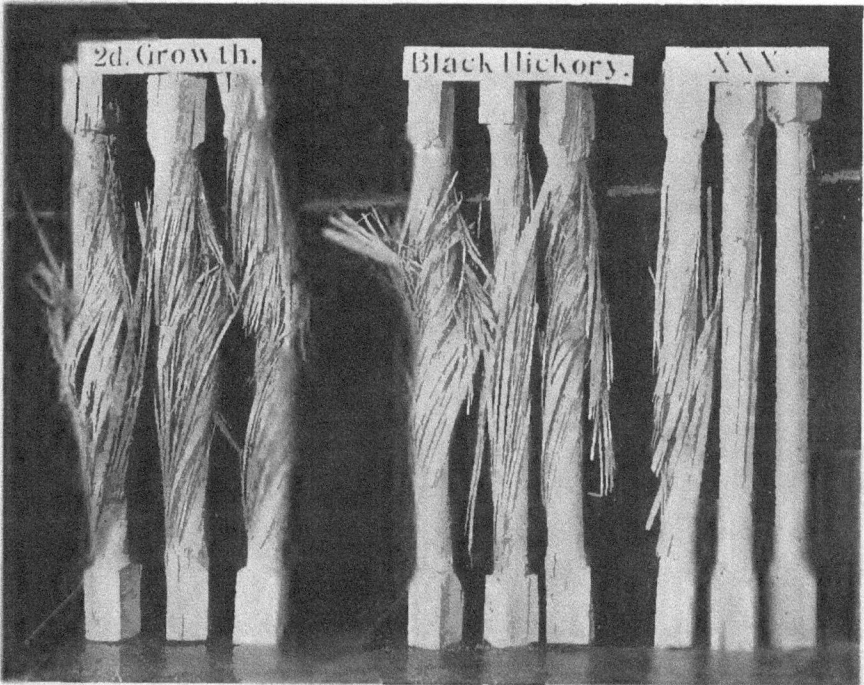

Photo by U. S. Forest Service.

Fig. 20.—Effect of torsion on different grades of hickory.

wearing ability of wood and is an important consideration in the use of wood for floors, paving blocks, bearings, and rollers. While resistance to indentation is dependent mostly upon the density of the wood, the wearing qualities may be governed by other factors such as toughness, and the size, cohesion, and arrangement of the fibres. In use for floors, some woods tend to compact and wear smooth, while others become splintery and rough. This feature is affected to some extent by the manner in which the wood is sawed; thus edge-grain pine flooring is much better than flat-sawn for uniformity of wear.

TABLE XII

HARDNESS OF 32 WOODS IN GREEN CONDITION, AS INDICATED BY THE LOAD
REQUIRED TO IMBED A 0.444-INCH STEEL BALL TO ONE-HALF ITS
DIAMETER

(Forest Service Cir. 213)

COMMON NAME OF SPECIES	Average	End surface	Radial surface	Tangential surface
	Pounds	*Pounds*	*Pounds*	*Pounds*
Hardwoods				
1 Osage orange......	1,971	1,838	2,312	1,762
2 Honey locust......	1,851	1,862	1,860	1,832
3 Swamp white oak ..	1,174	1,205	1,217	1,099
4 White oak.........	1,164	1,183	1,163	1,147
5 Post oak..........	1,099	1,139	1,068	1,081
6 Black oak.........	1,069	1,093	1,083	1,031
7 Red oak..........	1,043	1,107	1,020	1,002
8 White ash........	1,046	1,121	1,000	1,017
9 Beech............	942	1,012	897	918
10 Sugar maple.......	937	992	918	901
11 Rock elm..	910	954	883	893
12 Hackberry........	799	829	795	773
13 Slippery elm... ..	788	919	757	687
14 Yellow birch.. ...	778	827	768	739
15 Tupelo.....	738	814	666	733
16 Red maple.. ...	671	766	621	626
17 Sycamore... ...	608	664	560	599
18 Black ash......	551	565	542	546
19 White elm.....	496	536	456	497
20 Basswood.......	239	273	226	217
Conifers				
1 Longleaf pine.. .	532	574	502	521
2 Douglas fir....	410	415	399	416
3 Bald cypress.. .	390	460	355	354
4 Hemlock......	384	463	354	334
5 Tamarack.....	384	401	380	370
6 Red pine..........	347	355	345	340
7 White fir.........	346	381	322	334
8 Western yellow pine	328	334	307	342
9 Lodgepole pine.....	318	316	318	319
10 White pine........	299	304	294	299
11 Engelmann spruce..	266	272	253	274
12 Alpine fir.........	241	284	203	235

NOTE.—Black locust and hickory are not included in this table, but their
position would be near the head of the list.

Tests for either form of hardness are of comparative value only. Tests for indentation are commonly made by penetrations of the material with a steel punch or ball.* Tests for abrasion are made by wearing down wood with sandpaper or by means of a sand blast.

CLEAVABILITY

Cleavability is the term used to denote the facility with which wood is split. A splitting stress is one in which the forces act normally like a wedge. (See Fig. 21.) The plane of cleavage is parallel to the grain, either radially or tangentially.

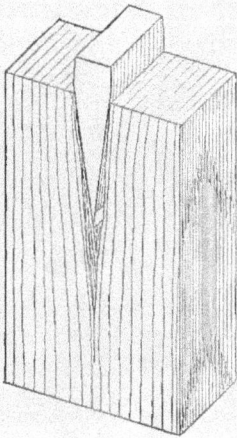

FIG. 21.—Cleavage of highly elastic wood. The cleft runs far ahead of the wedge.

This property of wood is very important in certain uses such as firewood, fence rails, billets, and squares. Resistance to splitting or low cleavability is desirable where wood must hold nails or screws, as in box-making. Wood usually splits more readily along the radius than parallel to the growth rings though exceptions occur, as in the case of cross grain.

Splitting involves transverse tension, but only a portion of the fibres are under stress at a time. A wood of little stiffness and strong cohesion across the grain is difficult to split, while one with great stiffness, such as longleaf pine, is easily split. The form of the grain and the presence of knots greatly affect this quality.

---

* See articles by Gabriel Janka listed in bibliography, pages 151–152.

TABLE XIII

CLEAVAGE STRENGTH OF SMALL CLEAR PIECES OF 32 WOODS IN GREEN
CONDITION

(Forest Service Cir. 213)

COMMON NAME OF SPECIES	When surface of failure is radial	When surface of failure is tangential
	Lbs. per sq. in. of width	*Lbs. per sq. in. of width*
Hardwoods		
Ash, black...............	275	260
white................	333	346
Basswood................	130	168
Beech...................	339	527
Birch, yellow............	294	287
Elm, slippery............	401	424
white............	210	270
Hackberry...............	422	436
Locust, honey...........	552	610
Maple, red..............	297	330
sugar.............	376	513
Oak, post...............	354	487
red................	380	470
swamp white.......	428	536
white..............	382	457
yellow.............	379	470
Sycamore...............	265	425
Tupelo..................	277	380
Conifers		
Arborvitæ...............	148	139
Cypress, bald...........	167	154
Fir, alpine..............	130	133
Douglas............	139	127
white..............	145	187
Hemlock...............	168	151
Pine, lodgepole.........	142	140
longleaf...........	187	180
red................	161	154
sugar..............	168	189
western yellow.....	162	187
white..............	144	160
Spruce, Engelmann......	110	135
Tamarack...............	167	159

PART II

FACTORS AFFECTING THE MECHANICAL PROPERTIES OF WOOD

INTRODUCTION

WCOD is an organic product—a structure of infinite variation of detail and design.* It is on this account that no two woods are alike—in reality no two specimens from the same log are identical. There are certain properties that characterize each species, but they are subject to considerable variation. Oak, for example, is considered hard, heavy, and strong, but some pieces, even of the same species of oak, are much harder, heavier, and stronger than others. With hickory are associated the properties of great strength, toughness, and resilience, but some pieces are comparatively weak and brash and ill-suited for the exacting demands for which good hickory is peculiarly adapted.

It follows that no definite value can be assigned to the properties of any wood and that tables giving average results of tests may not be directly applicable to any individual stick. With sufficient knowledge of the intrinsic factors affecting the results it becomes possible to infer from the appearance of material its probable variation from the average. As yet too little is known of the relation of structure and chemical composition to the mechanical and physical properties to permit more than general conclusions.

RATE OF GROWTH

To understand the effect of variations in the rate of growth it is first necessary to know how wood is formed. A tree increases in diameter by the formation, between the old wood and the inner bark, of new woody layers which envelop the entire stem, living

---

* For details regarding the structure of wood see Record, Samuel J.: Identification of the economic woods of the United States. New York, John Wiley & Sons, 1912.

branches, and roots. Under ordinary conditions one layer is formed each year and in cross section as on the end of a log they appear as rings—often spoken of as *annual rings*. These growth layers are made up of wood cells of various kinds, but for the most part fibrous. In timbers like pine, spruce, hemlock, and other coniferous or softwood species the wood cells are mostly of one kind, and as a result the material is much more uniform in structure than that of most hardwoods. (See Frontispiece.) There are no vessels or pores in coniferous wood such as one sees so prominently in oak and ash, for example. (See Fig. 22.)

The structure of the hardwoods is more complex. They are more or less filled with vessels, in some cases (oak, chestnut, ash) quite large and distinct, in others (buckeye, poplar, gum) too small to be seen plainly without a small hand lens. In discussing such woods it is customary to divide them into two large classes —*ring-porous* and *diffuse-porous*. (See Fig. 22.) In ring-porous species, such as oak, chestnut, ash, black locust, catalpa, mulberry, hickory, and elm, the larger vessels or pores (as cross sections of vessels are called) become localized in one part of the growth ring, thus forming a region of more or less open and porous tissue. The rest of the ring is made up of smaller vessels and a much greater proportion of wood fibres. These fibres are the elements which give strength and toughness to wood, while the vessels are a source of weakness.

In diffuse-porous woods the pores are scattered throughout the growth ring instead of being collected in a band or row. Examples of this kind of wood are gum, yellow poplar, birch, maple, cottonwood, basswood, buckeye, and willow. Some species, such as walnut and cherry, are on the border between the two classes, forming a sort of intermediate group.

If one examines the smoothly cut end of a stick of almost any kind of wood, he will note that each growth ring is made up of two more or less well-defined parts. That originally nearest the centre of the tree is more open textured and almost invariably lighter in color than that near the outer portion of the ring. The inner portion was formed early in the season, when growth was comparatively rapid and is known as *early wood* (also spring wood); the outer portion is the *late wood*, being produced in the summer or early fall. In soft pines there is not much contrast in the different parts of the ring, and as a result the wood is very uniform in texture and is easy to work. In hard pine, on the

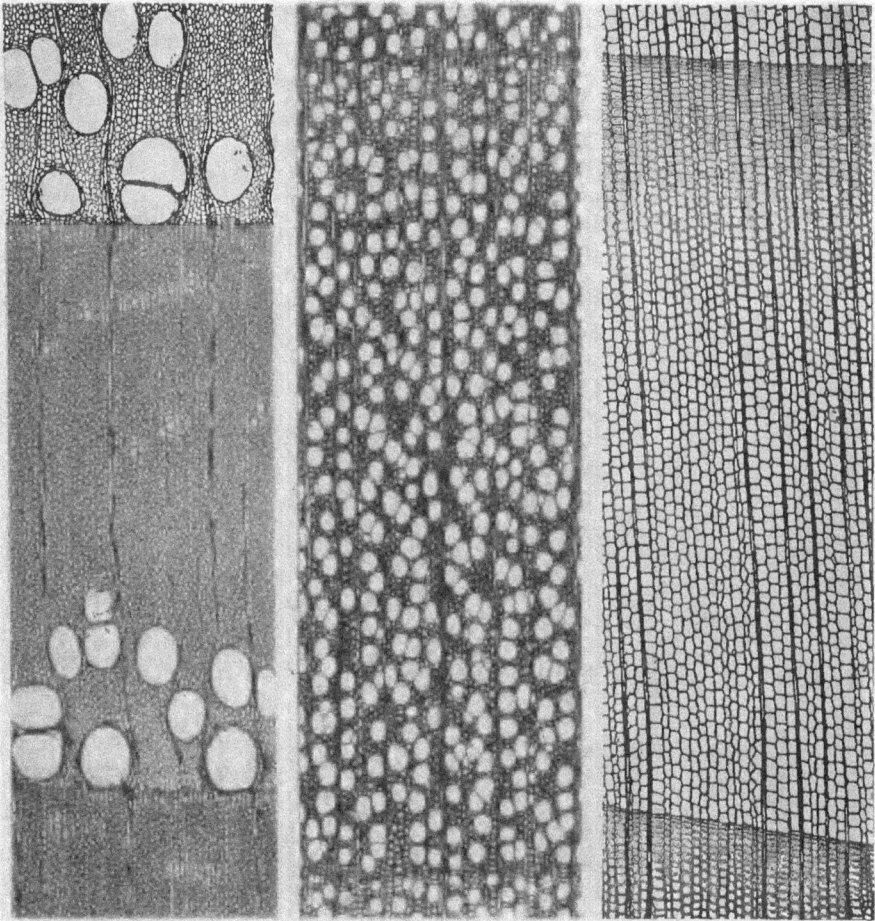

Photomicrographs by the author.

Fig. 22.—Cross sections of a ring-porous hardwood (white ash), a diffuse-porous hardwood (red gum), and a non-porous or coniferous wood (eastern hemlock). ×30.

other hand, the late wood is very dense and is deep-colored, presenting a very decided contrast to the soft, straw-colored early wood. (See Fig. 23.) In ring-porous woods each season's growth is always well defined, because the large pores of the spring abut on the denser tissue of the fall before. In the diffuse-porous, the demarcation between rings is not always so clear and in not a few cases is almost, if not entirely, invisible to the unaided eye. (See Fig. 22.)

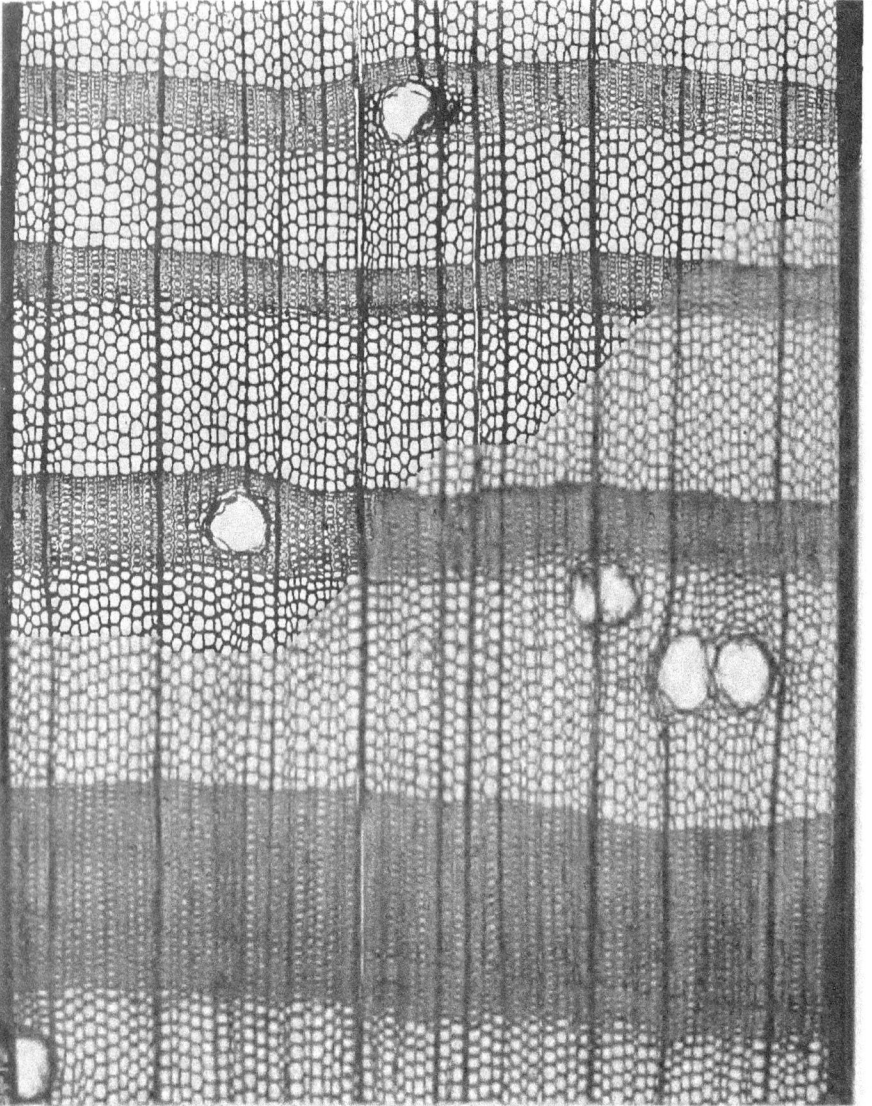

Photomicrograph by U. S. Forest Service.

FIG. 23.—Cross section of longleaf pine showing several growth rings with variations in the width of the dark-colored late wood. Seven resin ducts are visible. ×3).

If one compares a heavy piece of pine with a light specimen it will be seen at once that the heavier one contains a larger proportion of late wood than the other, and is therefore considerably darker. The late wood of all species is denser than that formed early in the season, hence the greater the proportion of late wood the greater the density and strength. When examined under a microscope the cells of the late wood are seen to be very thick-walled and with very small cavities, while those formed first in the season have thin walls and large cavities. The strength is in the walls, not the cavities. In choosing a piece of pine where strength or stiffness is the important consideration, the principal thing to observe is the comparative amounts of early and late wood. The width of ring, that is, the number per inch, is not nearly so important as the proportion of the late wood in the ring.

It is not only the proportion of late wood, but also its quality, that counts. In specimens that show a very large proportion of late wood it may be noticeably more porous and weigh consid-erably less than the late wood in pieces that contain but little. One can judge comparative density, and therefore to some extent weight and strength, by visual inspection.

The conclusions of the U. S. Forest Service regarding the effect of rate of growth on the properties of Douglas fir are sum-marized as follows:

" 1. In general, rapidly grown wood (less than eight rings per inch) is relatively weak. A study of the individual tests upon which the average points are based shows, however, that when it is not associated with light weight and a small proportion of summer wood, rapid growth is not indicative of weak wood.

" 2. An average rate of growth, indicated by from 12 to 16 rings per inch, seems to produce the best material.

" 3. In rates of growths lower than 16 rings per inch, the average strength of the material decreases, apparently approaching a uniform condition above 24 rings per inch. In such slow rates of growth the texture of the wood is very uniform, and naturally there is little variation in weight or strength.

" An analysis of tests on large beams was made to ascertain if average rate of growth has any relation to the mechanical prop-erties of the beams. The analysis indicated conclusively that there was no such relation. Average rate of growth [without con-sideration also of density], therefore, has little significance in

grading structural timber." *   This is because of the wide variation in the percentage of late wood in different parts of the
cross section.

Experiments seem to indicate that for most species there is a
rate of growth which, in general, is associated with the greatest
strength, especially in small specimens.   For eight conifers it is as
follows: †

	Rings per inch
Douglas fir	24
Shortleaf pine	12
Loblolly pine	6
Western larch	18
Western hemlock	14
Tamarack	20
Norway pine	18
Redwood	30

No satisfactory explanation can as yet be given for the real
causes underlying the formation of early and late wood.   Several
factors may be involved.   In conifers, at least, rate of growth alone
does not determine the proportion of the two portions of the ring,
for in some cases the wood of slow growth is very hard and heavy,
while in others the opposite is true.   The quality of the site where
the tree grows undoubtedly affects the character of the wood
formed, though it is not possible to formulate a rule governing
it.   In general, however, it may be said that where strength or
ease of working is essential, woods of moderate to slow growth
should be chosen.   But in choosing a particular specimen it is
not the width of ring, but the proportion and character of the
late wood which should govern.

In the case of the ring-porous hardwoods there seems to exist
a pretty definite relation between the rate of growth of timber
and its properties.   This may be briefly summed up in the general
statement that the more rapid the growth or the wider the rings
of growth, the heavier, harder, stronger, and stiffer the wood.
This, it must be remembered, applies only to ring-porous woods
such as oak, ash, hickory, and others of the same group, and is,
of course, subject to some exceptions and limitations.

In ring-porous woods of good growth it is usually the middle

---

* Bul. 88: Properties and uses of Douglas fir, p. 29.
† Bul. 108, U. S. Forest Service: Tests of structural timbers, p. 37.

portion of the ring in which the thick-walled, strength-giving
fibres are most abundant. As the breadth of ring diminishes, this
middle portion is reduced so that very slow growth produces comparatively light, porous wood composed of thin-walled vessels and
wood parenchyma. In good oak these large vessels of the early
wood occupy from 6 to 10 per cent of the volume of the log, while
in inferior material they may make up 25 per cent or more. The
late wood of good oak, except for radial grayish patches of small
pores, is dark colored and firm, and consists of thick-walled fibres
which form one-half or more of the wood. In inferior oak, such
fibre areas are much reduced both in quantity and quality. Such
variation is very largely the result of rate of growth.

Wide-ringed wood is often called " second-growth," because
the growth of the young timber in open stands after the old trees
have been removed is more rapid than in trees in the forest, and
in the manufacture of articles where strength is an important
consideration such " second-growth " hardwood material is preferred. This is particularly the case in the choice of hickory for
handles and spokes. Here not only strength, but toughness and
resilience are important. The results of a series of tests on
hickory by the U. S. Forest Service show that " the work or
shock-resisting ability is greatest in wide-ringed wood that has
from 5 to 14 rings per inch, is fairly constant from 14 to 38 rings,
and decreases rapidly from 38 to 47 rings. The strength at maximum load is not so great with the most rapid-growing wood; it is
maximum with from 14 to 20 rings per inch, and again becomes
less as the wood becomes more closely ringed. The natural deduc-
tion is that wood of first-class mechanical value shows from 5 to
20 rings per inch and that slower growth yields poorer stock.
Thus the inspector or buyer of hickory should discriminate against
timber that has more than 20 rings per inch. Exceptions exist,
however, in the case of normal growth upon dry situations, in
which the slow-growing material may be strong and tough." *

The effect of rate of growth on the qualities of chestnut wood
is summarized by the same authority as follows: " When the
rings are wide, the transition from spring wood to summer wood

* Bul. 80: The commercial hickories, pp. 48–50.

is gradual, while in the narrow rings the spring wood passes into summer wood abruptly. The width of the spring wood changes but little with the width of the annual ring, so that the narrowing or broadening of the annual ring is always at the expense of the summer wood. The narrow vessels of the summer wood make it richer in wood substance than the spring wood composed of wide vessels. Therefore, rapid-growing specimens with wide rings have more wood substance than slow-growing trees with narrow rings. Since the more the wood substance the greater the weight, and the greater the weight the stronger the wood, chestnuts with wide rings must have stronger wood than chestnuts with narrow rings. This agrees with the accepted view that sprouts (which always have wide rings) yield better and stronger wood than seedling chestnuts, which grow more slowly in diameter." *

In diffuse-porous woods, as has been stated, the vessels or pores are scattered throughout the ring instead of collected in the early wood. The effect of rate of growth is, therefore, not the same as in the ring-porous woods, approaching more nearly the conditions in the conifers. In general it may be stated that such woods of medium growth afford stronger material than when very rapidly or very slowly grown. In many uses of wood, strength is not the main consideration. If ease of working is prized, wood should be chosen with regard to its uniformity of texture and straightness of grain, which will in most cases occur when there is little contrast between the late wood of one season's growth and the early wood of the next.

HEARTWOOD AND SAPWOOD

Examination of the end of a log of many species reveals a darker-colored inner portion—the *heartwood*, surrounded by a lighter-colored zone—the *sapwood*. In some instances this distinction in color is very marked; in others, the contrast is slight, so that it is not always easy to tell where one leaves off and the other begins. The color of fresh sapwood is always light, sometimes pure white, but more often with a decided tinge of green or brown.

---

* Bul. 53: Chestnut in southern Maryland, pp. 20–21.

Sapwood is comparatively new wood. There is a time in the early history of every tree when its wood is all sapwood. Its principal functions are to conduct water from the roots to the leaves and to store up and give back according to the season the food prepared in the leaves. The more leaves a tree bears and the more thrifty its growth, the larger the volume of sapwood required, hence trees making rapid growth in the open have thicker sapwood for their size than trees of the same species growing in dense forests. Sometimes trees grown in the open may become of considerable size, a foot or more in diameter, before any heartwood begins to form, for example, in second-growth hickory, or field-grown white and loblolly pines.

As a tree increases in age and diameter an inner portion of the sapwood becomes inactive and finally ceases to function. This inert or dead portion is called heartwood, deriving its name solely from its position and not from any vital importance to the tree, as is shown by the fact that a tree can thrive with its heart completely decayed. Some species begin to form heartwood very early in life, while in others the change comes slowly. Thin sapwood is characteristic of such trees as chestnut, black locust, mulberry, Osage orange, and sassafras, while in maple, ash, gum, hickory, hackberry, beech, and loblolly pine, thick sapwood is the rule.

There is no definite relation between the annual rings of growth and the amount of sapwood. Within the same species the cross-sectional area of the sapwood is roughly proportional to the size of the crown of the tree. If the rings are narrow, more of them are required than where they are wide. As the tree gets larger, the sapwood must necessarily become thinner or increase materially in volume. Sapwood is thicker in the upper portion of the trunk of a tree than near the base, because the age and the diameter of the upper sections are less.

When a tree is very young it is covered with limbs almost, if not entirely, to the ground, but as it grows older some or all of them will eventually die and be broken off. Subsequent growth of wood may completely conceal the stubs which, however, will remain as knots. No matter how smooth and clear a log is on the outside, it is more or less knotty near the middle. Consequently the sapwood of an old tree, and particularly of a forest-grown tree, will be freer from knots than the heartwood. Since in most uses of wood, knots are defects that weaken the timber and interfere with its ease of working and other properties, it follows that sapwood, because of its position in the tree, may have certain advantages over heartwood.

It is really remarkable that the inner heartwood of old trees remains as sound as it usually does, since in many cases it is hundreds of years, and in a few instances thousands of years, old. Every broken limb or root, or deep wound from fire, insects, or falling timber, may afford an entrance for decay, which, once started, may penetrate to all parts of the trunk. The larvæ of many insects bore into the trees and their tunnels remain indefinitely as sources of weakness. Whatever advantages, however, that sapwood may have in this connection are due solely to its relative age and position.

If a tree grows all its life in the open and the conditions of soil and site remain unchanged, it will make its most rapid growth in youth, and gradually decline. The annual rings of growth are for many years quite wide, but later they become narrower and narrower. Since each succeeding ring is laid down on the outside of the wood previously formed, it follows that unless a tree materially increases its production of wood from year to year, the rings must necessarily become thinner. As a tree reaches maturity its crown becomes more open and the annual wood production is lessened, thereby reducing still more the width of the growth rings. In the case of forest-grown trees so much depends upon the competition of the trees in their struggle for light and nourishment that periods of rapid and slow growth may alternate. Some trees, such as southern oaks, maintain the same width of ring for hundreds of years. Upon the whole, however, as a tree gets larger in diameter the width of the growth rings decreases.

It is evident that there may be decided differences in the grain of heartwood and sapwood cut from a large tree, particularly one that is overmature. The relationship between width of growth rings and the mechanical properties of wood is discussed under Rate of Growth. In this connection, however, it may be stated that as a general rule the wood laid on late in the life of a tree is softer, lighter, weaker, and more even-textured than that produced earlier. It follows that in a large log the sapwood, because of the time in the life of the tree when it was grown, may be inferior in hardness, strength, and toughness to equally sound heartwood from the same log.

After exhaustive tests on a number of different woods the U. S. Forest Service concludes as follows: " Sapwood, except that from old, overmature trees, is as strong as heartwood, other things being equal, and so far as the mechanical properties go should not be regarded as a defect." * Careful inspection of the individual tests made in the investigation fails to reveal any relation between the proportion of sapwood and the breaking strength of timber.

In the study of the hickories the conclusion was: " There is an unfounded prejudice against the heartwood. Specifications place white hickory, or sapwood, in a higher grade than red hickory, or heartwood, though there is no inherent difference in strength. In fact, in the case of large and old hickory trees, the sapwood nearest the bark is comparatively weak, and the best wood is in the heart, though in young trees of thrifty growth the best wood is in the sap." † The results of tests from selected pieces lying side by side in the same tree, and also the average values for heartwood and sapwood in shipments of the commercial hickories without selection, show conclusively that " the transformation of sapwood into heartwood does not affect either the strength or toughness of the wood. . . . It is true, however, that sapwood is usually more free from latent defects than heartwood." ‡

Specifications for paving blocks often require that longleaf pine be 90 per cent heart. This is on the belief that sapwood is not only more subject to decay, but is also weaker than heartwood. In reality there is no sound basis for discrimination against sapwood on account of strength, provided other conditions are equal. It is true that sapwood will not resist decay as long as heartwood, if both are untreated with preservatives. It is especially so of woods with deep-colored heartwood, and is due to infiltrations of tannins, oils, and resins, which make the wood more or

---

* Bul. 108: Tests of structural timber, p. 35.
† Bul. 80: The commercial hickories, p. 50.
‡ Loc. cit.

less obnoxious to decay-producing fungi. If, however, the timbers are to be treated, sapwood is not a defect; in fact, because of the relative ease with which it can be impregnated with preservatives it may be made more desirable than heartwood.*

In specifications for structural timbers reference is sometimes made to " boxheart," meaning the inclusion of the pith or centre of the tree within a cross section of the timber. From numerous experiments it appears that the position of the pith does not bear any relation to the strength of the material. Since most season checks, however, are radial, the position of the pith may influence the resistance of a seasoned beam to horizontal shear, being greatest when the pith is located in the middle half of the section.†

WEIGHT, DENSITY, AND SPECIFIC GRAVITY

From data obtained from a large number of tests on the strength of different woods it appears that, other things being equal, the crushing strength parallel to the grain, fibre stress at elastic limit in bending, and shearing strength along the grain of wood vary in direct proportion to the weight of dry wood per unit of volume when green. Other strength values follow different laws. The hardness varies in a slightly greater ratio than the square of the density. The work to the breaking point increases even more rapidly than the cube of density. The modulus of rupture in bending lies between the first power and the square of the density. This, of course, is true only in case the greater weight is due to increase in the amount of wood substance. A

---

* Although the factor of heart or sapwood does not influence the mechanical properties of the wood and there is usually no difference in structure observable under the microscope, nevertheless sapwood is generally decidedly different from heartwood in its physical properties. It dries better and more easily than heartwood, usually with less shrinkage and little checking or honeycombing. This is especially the case with the more refractory woods, such as white oaks and *Eucalyptus globulus* and *viminalis*. It is usually much more permeable to air, even in green wood, notably so in loblolly pine and even in white oak. As already stated, it is much more subject to decay. The sapwood of white oak may be impregnated with creosote with comparative ease, while the heartwood is practically impenetrable. These facts indicate a difference in its chemical nature.—H. D. Tiemann. .

† Bul. 108, U. S. Forest Service, p. 36.

wood heavy with resin or other infiltrated substance is not necessarily stronger than a similar specimen free from such materials.
If differences in weight are due to degree of seasoning, in other
words, to the relative amounts of water contained, the rules given
above will of course not hold, since strength increases with dryness.
But of given specimens of pine or of oak, for example, in the green
condition, the comparative strength may be inferred from the
weight. It is not permissible, however, to compare such widely
different woods as oak and pine on a basis of their weights.*

The weight of wood substance, that is, the material which com-
poses the walls of the fibres and other cells, is practically the same
in all species, whether pine, hickory, or cottonwood, being a little
greater than half again as heavy as water. It varies slightly
from beech sapwood, 1.50, to Douglas fir heartwood, 1.57, averaging
about 1.55 at 30° to 35° C., in terms of water at its greatest density
4° C. The reason any wood floats is that the air imprisoned in its
cavities buoys it up. When this is displaced by water the wood
becomes water-logged and sinks. Leaving out of consideration
infiltrated substances, the reason a cubic foot of one kind of dry
wood is heavier than that of another is because it contains a greater
amount of wood substance.

Density is merely the weight of a unit of volume, as 35 pounds
per cubic foot, or 0.56 grams per cubic centimetre. **Specific gravity**
or relative density is the ratio of the density of any material to
the density of distilled water at 4° C. (39.2° F.). A cubic foot of
distilled water at 4° C. weighs 62.43 pounds. Hence the specific
gravity of a piece of wood with a density of 35 pounds is $\dfrac{35}{62.43}$
= 0.561. To find the weight per cubic foot when the specific gravity is given, simply múltiply by 62.43. Thus, 0.561 × 62.43 = 35.
In the metric system, since the weight of a cubic centimetre of
pure water is one gram, the density in grams per cubic centimetre has the same numerical value as the specific gravity.

Since the amount of water in wood is extremely variable it
usually is not satisfactory to refer to the density of green wood.

---

* The oaks for some unknown reason fall below the normal strength for
weight, whereas the hickories rise above. Certain other woods also are somewhat exceptional to the normal relation of strength and density.

TABLE XIV

SPECIFIC GRAVITY, AND SHRINKAGE OF 51 AMERICAN WOODS
(Forest Service Cir. 213)

COMMON NAME OF SPECIES	Moisture content	Specific gravity oven-dry, based on		Shrinkage from green to oven-dry condition		
		Volume when green	Volume when oven-dry	In volume	Radial	Tangential
Hardwoods	*Per cent*			*Per cent*	*Per cent*	*Per cent*
Ash, black.......	77	0.466	.....	....	...	....
white.......	38	.550	0.640	12.6	4.3	6.4
"	47	.516	.590	11.7	...	....
Basswood.......	110	.315	.374	14.5	6.2	8.4
Beech..........	61	.556	.669	16.5	4.6	10.5
Birch, yellow.....	72	.545	.661	17.0	7.9	9.0
Elm, rock.......	46	.578	.....	....	...	....
slippery....	57	.541	.639	15.5	5.1	9.9
white......	66	.430	.....	....	...	....
Gum, red........	71	.434	.....	....	...	....
Hackberry.......	50	.504	.576	14.0	4.2	8.9
Hickory,						
big shellbark..	64	.601	.....	17.6	7.4	11.2
" " ..	55	.666	.....	20.9	7.9	14.2
bitternut......	65	.624	.....	....	...	....
mockernut.....	64	.606	.....	16.5	6.9	10.4
"	57	.662	.....	18.9	8.4	11.4
"	48	.666	.....	....	...	....
nutmeg........	76	.558	.....	....	...	....
pignut........	59	.627	.....	15.0	5.6	9.8
"	54	.667	.....	15.3	6.3	9.5
"	55	.667	.....	16.9	6.8	10.9
"	52	.667	.....	21.2	8.5	13.8
shagbark......	65	.608	.....	16.0	6.5	10.2
"	58	.646	.....	18.4	7.9	11.4
"	64	.617	.....	....	...	....
"	60	.653	.....	15.5	6.5	9.7
water........	74	.630	.....	....	...	....
Locust, honey....	53	.695	.759	8.6	...	....
Maple, red......	69	.512	.....	....	...	....
sugar.....	57	.546	.643	14.3	4.9	9.1
"	56	.577	.....	....	...	....
Oak, post........	64	.590	.732	16.0	5.7	10.6
red........	80	.568	.660	13.1	3.7	8.3
swamp white	74	.637	.792	17.7	5.5	10.6
tanbark....	88	.585	.....	....	...	....
white......	58	.594	.704	15.8	6.2	8.3
"	62	.603	.696	14.3	4.9	9.0
"	78	.600	.708	16.0	4.8	9.2
yellow......	77	.573	.669	14.2	4.5	9.7
"	80	.550	.....	....	...	....
Osage orange.....	31	.761	.838	8.9	...	....
Sycamore.......	81	.454	.526	13.5	5.0	7.3
Tupelo.... ...	121	.475	.545	12.4	4.4	7.9

TABLE XIV.—*Continued*

COMMON NAME OF SPECIES	Moisture content	Specific gravity oven-dry, based on		Shrinkage from green to oven-dry condition		
		Volume when green	Volume when oven-dry	In volume	Radial	Tangential
Conifers	*Per cent*			*Per cent*	*Per cent*	*Per cent*
Arborvitæ........	55	.293	.315	7.0	2.1	4.9
Cedar, incense....	80	.363	.....	....	...	....
Cypress, bald....	79	.452	.513	11.5	3.8	6.0
Fir, alpine.......	47	.306	.321	9.0	2.5	7.1
amabilis.....	117	.383	.....	....	...	....
Douglas.....	32	.418	.458	10.9	3.7	6.6
white.......	156	.350	.437	10.2	3.4	7.0
Hemlock (east.)..	129	.340	.394	9.2	2.3	5.0
Pine, lodgepole...	44	.370	.415	11.3	4.2	7.1
" ...	58	.371	.407	10.1	3.6	5.9
longleaf....	63	.528	.599	12.8	6.0	7.6
red or Nor..	54	.440	.507	11.5	4.5	7.2
shortleaf....	52	.447	.....	....	...	....
sugar......	123	.360	.386	8.4	2.9	5.6
west. yellow	98	.353	.395	9.2	4.1	6.4
" "	125	.377	.433	11.5	4.3	7.3
" "	93	.391	.435	9.9	3.8	5.8
white.......	74	.363	.391	7.8	2.2	5.9
Redwood........	81	.334	....	....	...	....
"	69	.366	.....	....	...	....
Spruce, Engelmann....	45	.325	359	10.5	3.7	6.9
"	156	.299	.335	10.3	3.0	6.2
red..........	31	.396	.....	....	...	....
white........	41	.318	.....	....	...	....
Tamarack.......	52	.491	.558	13.6	3.7	7.4

For scientific purposes the density of " oven-dry " wood is used; that is, the wood is dried in an oven at a temperature of 100° C. (212° F.) until a constant weight is attained. For commercial purposes the weight or density of air-dry or " shipping-dry " wood is used. This is usually expressed in pounds per thousand board feet, a board foot being considered as one-twelfth of a cubic foot.

Wood shrinks greatly in drying from the green to the oven-dry condition. (See Table XIV.) Consequently a block of wood measuring a cubic foot when green will measure considerably less when oven-dry. It follows that the density of oven-dry wood does not represent the weight of the dry wood substance in a cubic foot of green wood. In other words, it is not the weight of a cubic foot of green wood minus the weight of the water which it contains. Since the latter is often a more convenient figure to use and much easier to obtain than the weight of oven-dry wood, it is commonly expressed in tables of " specific gravity or density of dry wood."

This weight divided by 62.43 gives the specific gravity per green volume. It is purely a fictitious quantity. To convert this figure into actual density or specific gravity of the dry wood, it is necessary to know the amount of shrinkage in volume. If S is the percentage of shrinkage from the green to the oven-dry condition, based on the green volume; D, the density of the dry wood per cubic foot while green; and d the actual density of oven-dry wood, then $\dfrac{D}{1 - .0S} = d$.

This relation becomes clearer from the following analysis: Taking V and W as the volume and weight, respectively, when green, and v and w as the corresponding volume and weight when oven-dry, then, $d = \dfrac{w}{v}$; $D = \dfrac{W}{V}$; $S = \dfrac{V - v}{V} \times 100$, and $s = \dfrac{V - v}{v} \times 100$, in which S is the percentage of shrinkage from the green to the oven-dry condition, based on the green volume, and s the same based on the oven-dry volume.

In tables of specific gravity or density of wood it should always be stated whether the dry weight per unit of volume when green or the dry weight per unit of volume when dry is intended, since the shrinkage in volume may vary from 6 to 50 per cent, though in conifers it is usually about 10 per cent, and in hardwoods nearer 15 per cent. (See Table XIV.)

COLOR

In species which show a distinct difference between heartwood and sapwood the natural color of heartwood is invariably darker than that of the sapwood, and very frequently the contrast is conspicuous. This is produced by deposits in the heartwood of various materials resulting from the process of growth, increased possibly by oxidation and other chemical changes, which usually have little or no appreciable effect on the mechanical properties of

the wood. (See Heartwood and Sapwood.) Some experiments *
on very resinous longleaf pine specimens, however, indicate an increase in strength. This is due to the resin which increases the strength when dry. Spruce impregnated with crude resin and dried is greatly increased in strength thereby.

Since the late wood of a growth ring is usually darker in color than the early wood, this fact may be used in judging the density, and therefore the hardness and strength of the material. This is particularly the case with coniferous woods. In ring-porous woods the vessels of the early wood not infrequently appear on a finished surface as darker than the denser late wood, though on cross sections of heartwood the reverse is commonly true. Except in the manner just stated the color of wood is no indication of strength.

Abnormal discoloration of wood often denotes a diseased condition, indicating unsoundness. The black check in western hemlock is the result of insect attacks.† The reddish-brown streaks so common in hickory and certain other woods are mostly the result of injury by birds.‡ The discoloration is merely an indication of an injury, and in all probability does not of itself affect the properties of the wood. Certain rot-producing fungi impart to wood characteristic colors which thus become criterions of weakness. Ordinary sap-staining is due to fungous growth, but does not necessarily produce a weakening effect.§

CROSS GRAIN

Cross grain is a very common defect in timber. One form of it is produced in lumber by the method of sawing and has no reference to the natural arrangement of the wood elements. Thus

---

* Bul. 70, U. S. Forest Service, p. 92; also p. 126, appendix.

† See Burke, H. E.: Black check in western hemlock. Cir. No. 61, U. S. Bu. Entomology, 1905.

‡ See McAtee, W. L.: Woodpeckers in relation to trees and wood products. Bul. No. 39, U. S. Biol. Survey, 1911.

§ See Von Schrenck, Hermann: The "bluing" and the "red rot" of the western yellow pine, with special reference to the Black Hills forest reserve. Bul. No. 36, U. S. Bu. Plant Industry, Washington, 1903, pp. 13–14.

Weiss, Howard, and Barnum, Charles T.: The prevention of sapstain in lumber. Cir. 192, U. S. Forest Service, Washington, 1911, pp. 16–17.

if the plane of the saw is not approximately parallel to the axis of the log the grain of the lumber cut is not parallel to the edges and is termed diagonal. This is likely to occur where the logs have considerable taper, and in this case may be produced if sawed parallel to the axis of growth instead of parallel to the growth rings.

Lumber and timber with diagonal grain is always weaker than straight-grained material, the extent of the defect varying with the degree of the angle the fibres make with the axis of the stick. In the vicinity of large knots the grain is likely to be cross. The defect is most serious where wood is subjected to flexure, as in beams.

Spiral grain is a very common defect in a tree, and when excessive renders the timber valueless for use except in the round. It is produced by the arrangement of the wood fibres in a spiral direction about the axis instead of exactly vertical. Timber with spiral grain is also known as " torse wood." Spiral grain usually cannot be detected by casual inspection of a stick, since it does not show in the so-called visible grain of the wood, by which is commonly meant a sectional view of the annual rings of growth cut longitudinally. It is accordingly very easy to allow spiral-grained material to pass inspection, thereby introducing an element of weakness in a structure.

There are methods for readily detecting spiral grain. The simplest is that of splitting a small piece radially. It is necessary, of course, that the split be radial, that is, in a plane passing through the axis of the log, and not tangentially. In the latter case it is quite probable that the wood would split straight, the line of cleavage being between the growth rings.

In inspection, the elements to examine are the rays. In the case of oak and certain other hardwoods these rays are so large that they are readily seen not only on a radial surface, but on the tangential as well. On the former they appear as flakes, on the latter as short lines. Since these rays are between the fibres it naturally follows that they will be vertical or inclined according as the tree is straight-grained or spiral-grained. While they are not conspicuous in the softwoods, they can be seen upon close scrutiny, and particularly so if a small hand magnifier is used.

When wood has begun to dry and check it is very easy to see

whether or not it is straight- or spiral-grained, since the checks will for the most part follow along the rays. If one examines a row of telephone poles, for example, he will probably find that most of them have checks running spirally around them. If boards were sawed from such a pole after it was badly checked they would fall to pieces of their own weight. The only way to get straight material would be to split it out.

It is for this reason that split billets and squares are stronger than most sawed material. The presence of the spiral grain has little, if any, effect on the timber when it is used in the round, but in sawed material the greater the pitch of the spiral the greater is the defect.

KNOTS

Knots are portions of branches included in the wood of the stem or larger branch. Branches originate as a rule from the central axis of a stem, and while living increase in size by the addition of annual woody layers which are a continuation of those of the stem. The included portion is irregularly conical in shape with the tip at the pith. The direction of the fibre is at right angles or oblique to the grain of the stem, thus producing local cross grain.

During the development of a tree most of the limbs, especially the lower ones, die, but persist for a time—often for years. Subsequent layers of growth of the stem are no longer intimately joined with the dead limb, but are laid around it. Hence dead branches produce knots which are nothing more than pegs in a hole, and likely to drop out after the tree has been sawed into lumber. In grading lumber and structural timber, knots are classified according to their form, size, soundness, and the firmness with which they are held in place.*

Knots materially affect checking and warping, ease in working, and cleavability of timber. They are defects which weaken timber and depreciate its value for structural purposes where strength is an important consideration. The weakening effect is much more

---

* See Standard classification of structural timber. Yearbook Am. Soc. for Testing Materials, 1913, pp. 300–303. Contains three plates showing standard defects.

serious where timber is subjected to bending and tension than where under compression. The extent to which knots affect the strength of a beam depends upon their position, size, number, direction of fibre, and condition. A knot on the upper side is compressed, while one on the lower side is subjected to tension. The knot, especially (as is often the case) if there is a season check in it, offers little resistance to this tensile stress. Small knots, however, may be so located in a beam along the neutral plane as actually to increase the strength by tending to prevent longitudinal shearing. Knots in a board or plank are least injurious when they extend through it at right angles to its broadest surface. Knots which occur near the ends of a beam do not weaken it. Sound knots which occur in the central portion one-fourth the height of the beam from either edge are not serious defects.

Extensive experiments by the U. S. Forest Service * indicate the following effects of knots on structural timbers:

(1) Knots do not materially influence the stiffness of structural timber.

(2) Only defects of the most serious character affect the elastic limit of beams. Stiffness and elastic strength are more dependent upon the quality of the wood fibre than upon defects in the beam.

(3) The effect of knots is to reduce the difference between the fibre stress at elastic limit and the modulus of rupture of beams. The breaking strength is very susceptible to defects.

(4) Sound knots do not weaken wood when subject to compression parallel to the grain.†

FROST SPLITS

A common defect in standing timber results from radial splits which extend inward from the periphery of the tree, and almost, if not always, near the base. It is most common in trees which split readily, and those with large rays and thin bark. The

---

* Bul. 108, pp. 52 *et seq.*

† Bul. 115, U. S. Forest Service: Mechanical properties of western hemlock, p. 20.

primary cause of the splitting is frost, and various theories have been advanced to explain the action.

R. Hartig * believes that freezing forces out a part of the imbibition water of the cell walls, thereby causing the wood to shrink, and if the interior layers have not yet been cooled, tangential strains arise which finally produce radial clefts.

Another theory holds that the water is not driven out of the cell walls, but that difference in temperature conditions of inner and outer layers is itself sufficient to set up the strains, resulting in splitting. An air temperature of 14° F. or less is considered necessary to produce frost splits.

A still more recent theory is that of Busse † who considers the mechanical action of the wind a very important factor. He observed: (a) Frost splits sometimes occur at higher temperatures than 14° F. (b) Most splits take place shortly before sunrise, i.e., at the time of lowest air and soil temperature; they are never heard to take place at noon, afternoon, or evening. (c) They always occur between two roots or between the collars of two roots. (d) They are most frequent in old, stout-rooted, broad-crowned trees; in younger stands it is always the stoutest members that are found with frost splits, while in quite young stands they are altogether absent. (e) Trees on wet sites are most liable to splits, due to difference in wood structure, just as difference in wood structure makes different species vary in this regard. (f) Frost splits are most numerous less than three feet above the ground.

When a tree is swayed by the wind the roots are counteracting forces, and the wood fibres are tested in tension and compression by the opposing forces; where the roots exercise tension stresses most effectively the effect of compression stresses is at a minimum; only where the pressure is in excess of the tension, i.e., between the roots, can a separation of the fibre result. Hence, when by frost a tension on the entire periphery is established, and the

---

* Hartig, R.: The diseases of trees (trans. by Somerville and Ward), London and New York, 1894, pp. 282–294.

† Busse, W.: Frost-, Ring- und Kernrisse. Forstwiss. Centralb., XXXII, 2, 1910, pp. 74–81.

wind localizes additional strains, failure occurs. The stronger the compression and tension, the severer the strains and the oftener failures occur. The occurrence of reports of frost splits on wind-still days is believed by Busse to be due to the opening of old frost splits where the tension produced by the frost alone is sufficient.

Frost splits may heal over temporarily, but usually open up again during the following winter. The presence of old splits is often indicated by a ridge of callous, the result of the cambium's effort to occlude the wound. Frost splits not only affect the value of lumber, but also afford an entrance into the living tree for disease and decay.

SHAKES, GALLS, PITCH POCKETS

Heart shake occurs in nearly all overmature timber, being more frequent in hardwoods (especially oak) than in conifers. In typical heart shake the centre of the bole shows indications of becoming hollow and radial clefts of varying size extend outward from the pith, being widest inward. It frequently affects only the butt log, but may extend to the entire bole and even the larger branches. It usually results from a shrinkage of the heart-wood due probably to chemical changes in the wood.

When it consists of a single cleft extending across the pith it is termed *simple heart shake*. Shake of this character in straight-grained trees affects only one or two central boards when cut into lumber, but in spiral-grained timber the damage is much greater. When shake consists of several radial clefts it is termed *star shake*. In some instances one or more of these clefts may extend nearly to the bark. In felled or converted timber clefts due to heart shake may be distinguished from seasoning cracks by the darker color of the exposed surfaces. Such clefts, however, tend to open up more and more as the timber seasons.

Cup or *ring shake* results from the pulling apart of two or more growth rings. It is one of the most serious defects to which sound timber is subject, as it seriously reduces the technical prop-erties of wood. It is very common in sycamore and in western larch, particularly in the butt portion. Its occurrence is most frequent at the junction of two growth layers of very unequal thickness. Consequently it is likely to occur in trees which have grown slowly for a time, then abruptly increased, due to improved conditions of light and food, as in thinning. Old timber is more subject to it than young trees. The damage is largely confined to the butt log. Cup shake is often associated with other forms of shake, and not infrequently shows traces of decay.

The causes of cup shake are uncertain. The swaying action of the wind may result in shearing apart the growth layers, especially in trees growing in exposed places. Frost may in some instances be responsible for cup shake or at least a contributing factor, although trees growing in regions free from frost often have ring shake. Shrinkage of the heartwood may be concentric as well as radial in its action, thus producing cup shake instead of, or in connection with, heart shake.

A local defect somewhat similar in effect to cup shake is known as *rind gall*. If the cambium layer is exposed by the removal of the entire bark or rind it will die. Subsequent growth over the damaged portion does not cohere with the wood previously formed by the old cambium. The defect resulting is termed rind gall. The most common causes of it are fire, gnawing, blazing, chipping, sun scald, lightning, and abrasions.

Heart break is a term applied to areas of compression failure along the grain found in occasional logs. Sometimes these breaks are invisible until the wood is manufactured into the finished article. The occurrence of this defect is mostly limited to the dense hardwoods, such as hickory and to heavy tropical species. It is the source of considerable loss in the fancy veneer industry, as the veneer from valuable logs so affected drops to pieces.

The cause of heart break is not positively known. It is highly probable, however, that when the tree is felled the trunk strikes across a rock or another log, and the impact causes actual failure in the log as in a beam.

Resin or *pitch pockets* are of common occurrence in the wood of larch, spruce, fir, and especially of longleaf and other hard pines. They are due to accumulations of resin in openings between adjacent layers of growth. They are more frequent in trees growing alone than in those of dense stands. The pockets are usually a few inches in greatest dimension and affect only one or two growth layers. They are hidden until exposed by the saw, rendering it impossible to cut lumber with reference to their position. Often several boards are damaged by a single pocket. In grading lumber, pitch pockets are classified as small, standard, and large, depending upon their width and length.

INSECT INJURIES *

The larvæ of many insects are destructive to wood. Some attack the wood of living trees, others only that of felled or converted material. Every hole breaks the continuity of the fibres and impairs the strength, and if there are very many of them the material may be ruined for all purposes where strength is required.

Some of the most common insects attacking the wood of living trees are the oak timber worm, the chestnut timber worm, carpenter worms, ambrosia beetles, the locust borer, turpentine beetles and turpentine borers, and the white pine weevil.

The insect injuries to forest products may be classed according to the stage of manufacture of the material. Thus round timber with the bark on, such as poles, posts, mine props, and sawlogs, is subject to serious damage by the same class of insects as those mentioned above, particularly by the round-headed borers, timber worms, and ambrosia beetles. Manufactured unseasoned products are subject to damage from ambrosia beetles and other wood borers. Seasoned hardwood lumber of all kinds, rough handles, wagon stock, etc., made partially or entirely of sapwood, are often reduced in value from 10 to 90 per cent by a class of insects known as powder-post beetles. Finished hardwood products such as handles, wagon, carriage and machinery stock, especially if ash or hickory, are often destroyed by the powder-post beetles. Construction timbers in buildings, bridges and trestles, cross-ties, poles, mine props, fence posts, etc., are sometimes seriously injured by wood-boring larvæ, termites, black ants, carpenter bees, and powder-post beetles, and sometimes reduced

---

* For detailed information regarding insect injuries, the reader is referred to the various publications of the U. S. Bureau of Entomology, Washington, D. C.

in value from 10 to 100 per cent. In tropical countries termites are a very serious pest in this respect.

MARINE WOOD-BORER INJURIES

Vast amounts of timber used for piles in wharves and other marine structures are constantly being destroyed or seriously injured by marine borers. Almost invariably they are confined to salt water, and all the woods commonly used for piling are subject to their attacks. There are two genera of mollusks, *Xylotrya* and *Teredo*, and three of crustaceans, *Limnoria*, *Chelura*, and *Sphæroma*, that do serious damage in many places along both the Atlantic and Pacific coasts.

These mollusks, which are popularly known as " shipworms," are much alike in structure and mode of life. They attack the exposed surface of the wood and immediately begin to bore. The tunnels, often as large as a lead pencil, extend usually in a longitudinal direction and follow a very irregular, tangled course. Hard woods are apparently penetrated as readily as soft woods, though in the same timber the softer parts are preferred. The food consists of infusoria and is not obtained from the wood substance. The sole object of boring into the wood is to obtain shelter.

Although shipworms can live in cold water they thrive best and are most destructive in warm water. The length of time required to destroy an average barked, unprotected pine pile on the Atlantic coast south from Chesapeake Bay and along the entire Pacific coast varies from but one to three years.

Of the crustacean borers, *Limnoria*, or the " wood louse, is the only one of great importance, although *Sphæroma* is reported destructive in places. *Limnoria* is about the size of a grain of rice and tunnels into the wood for both food and shelter. The galleries extend inward radially, side by side, in countless numbers, to the depth of about one-half inch. The thin wood partitions remaining are destroyed by wave action, so that a fresh surface is exposed to attack. Both hard and soft woods are damaged, but the rate is faster in the soft woods or softer portions of a wood.

Timbers seriously attacked by marine borers are badly weakened or completely destroyed. If the original strength of the material is to be preserved it is necessary to protect the wood from the borers. This is sometimes accomplished by proper injection of creosote oil, and more or less successfully by the use of various kinds of external coatings.* No treatment, however, has proved entirely satisfactory.

FUNGOUS INJURIES †

Fungi are responsible for almost all decay of wood. So far as known, all decay is produced by living organisms, either fungi or bacteria. Some species attack living trees, sometimes killing them, or making them hollow, or in the case of pecky cypress and incense cedar filling the wood with galleries like those of boring insects. A much larger variety work only in felled or dead wood, even after it is placed in buildings or manufactured articles. In any case the process of destruction is the same. The mycelial threads penetrate the walls of the cells in search of food, which they find either in the cell contents (starches, sugars, etc.), or in the cell wall itself. The breaking down of the cell walls through the chemical action of so-called " enzymes " secreted by the fungi follows, and the eventual product is a rotten, moist substance crumbling readily under the slightest pressure. Some species remove the ligneous matter and leave almost pure cellulose, which is white, like cotton; others dissolve the cellulose, leaving a brittle, dark brown mass of ligno-cellulose. Fungi (such as the bluing fungus) which merely stain wood usually do not affect its mechanical properties unless the attacks are excessive.

It is evident, then, that the action of rot-causing fungi is to decrease the strength of wood, rendering it unsound, brittle, and

---

* See Smith, C. Stowell: Preservation of piling against marine wood borers. Cir. 128, U. S. Forest Service, 1908, pp. 15.

† See Von Schrenck, H.: The decay of timber and methods of preventing it. Bul. 14, U. S. Bu. Plant Industry, Washington, D. C., 1902. Also Buls. 32, 114, 214, 266.

Meinecke, E. P.: Forest tree diseases common in California and Nevada, U. S. Forest Service, Washington, D. C., 1914.

Hartig, R.: The diseases of trees. London and New York, 1894.

dangerous to use. The most dangerous kinds are the so-called "dry-rot" fungi which work in many kinds of lumber after it is placed in the buildings. They are particularly to be dreaded because unseen, working as they do within the walls or inside of casings. Several serious wrecks of large buildings have been attributed to this cause. It is stated * that in the three years (1911–1913) more than $100,000 was required to repair damage due to dry rot.

Dry rot develops best at 75° F. and is said to be killed by a temperature of 110° F.† Fully 70 per cent humidity is necessary in the air in which a timber is surrounded for the growth of this fungus, and probably the wood must be quite near its fibre saturation condition. Nevertheless *Merulius lacrymans* (one of the most important species) has been found to live four years and eight months in a dry condition.‡ Thorough kiln-drying will kill this fungus, but will not prevent its redevelopment. Antiseptic treatment, such as creosoting, is the best prevention.

All fungi require moisture and air § for their growth. Deprived of either of these the fungus dies or ceases to develop. Just what degree of moisture in wood is necessary for the "dry-rot" fungus has not been determined, but it is evidently considerably above that of thoroughly air-dry timber, probably more than 15 per cent moisture. Hence the importance of free circulation of air about all timbers in a building.

Warmth is also conducive to the growth of fungi, the most favorable temperature being about 90° F. They cannot grow in extreme cold, although no degree of cold such as occurs naturally will kill them. On the other hand, high temperature will kill them, but the spores may survive even the boiling temperature. Mould fungus has been observed to develop rapidly at 130° F. in a dry kiln in moist air, a condition under which an animal cannot

---

* Dry rot in factory timbers, by Inspection Dept. Associated Factory Mutual Fire Insurance Cos., 31 Milk Street, Boston, 1913.

† Falck, Richard: Die Meruliusfaüle des Bauholzes, Hausschwammforschungen, 6. Heft., Jena, 1912.

‡ Mez, Carl: Der Hausschwamm. Dresden, 1908, p. 63.

§ A culture of fungus placed in a glass jar and the air pumped out ceases to grow, but will start again as soon as oxygen is admitted.

live more than a few minutes. This fungus was killed, however, at about 140° or 145° F.*

The fungus (*Endothia parasitica* And.) which causes the chestnut blight kills the trees by girdling them and has no direct effect upon the wood save possibly the four or five growth rings of the sapwood.†

PARASITIC PLANT INJURIES.‡

The most common of the higher parasitic plants damaging timber trees are mistletoes. Many species of deciduous trees are attacked by the common mistletoe (*Phoradendron flavescens*). It is very prevalent in the South and Southwest and when present in sufficient quantity does considerable damage. There is also a considerable number of smaller mistletoes belonging to the genus *Razoumofskya* (*Arceuthobium*) which are widely distributed throughout the country, and several of them are common on coniferous trees in the Rocky Mountains and along the Pacific coast.

One effect of the common mistletoe is the formation of large swellings or tumors. Often the entire tree may become stunted or distorted. The western mistletoe is most common on the branches, where it produces " witches' broom." It frequently attacks the trunk as well, and boards cut from such trees are filled with long, radial holes which seriously damage or destroy the value of the timber affected.

LOCALITY OF GROWTH

The data available regarding the effect of the locality of growth upon the properties of wood are not sufficient to warrant

---

* Experiments in kiln-drying *Eucalyptus* in Berkeley, U. S. Forest Service.

† See Anderson, Paul J.: The morphology and life history of the chestnut blight fungus. Bul. No. 7, Penna. Chestnut Tree Blight Com., Harrisburg, 1914, p. 17.

‡ See York, Harlan H.: The anatomy and some of the biological aspects of the "American mistletoe." Bul. 120, Sci. Ser. No. 13, Univ. of Texas, Austin, 1909.

Bray, Wm. L.: The mistletoe pest in the Southwest. Bul. 166, U. S. Bu. Plant Ind., Washington, 1910.

Meinecke, E. P.: Forest tree diseases common in California and Nevada. U. S. Forest Service, Washington, 1914, pp. 54–58.

definite conclusions. The subject has, however, been kept in mind in many of the U. S. Forest Service timber tests and the following quotations are assembled from various reports:

" In both the Cuban and longleaf pine the locality where grown appears to have but little influence on weight or strength, and there is no reason to believe that the longleaf pine from one State is better than that from any other, since such variations as are claimed can be found on any 40-acre lot of timber in any State. But with loblolly and still more with shortleaf this seems not to be the case. Being widely distributed over many localities different in soil and climate, the growth of the shortleaf pine seems materially influenced by location. The wood from the southern coast and gulf region and even Arkansas is generally heavier than the wood from localities farther north. Very light and fine-grained wood is seldom met near the southern limit of the range, while it is almost the rule in Missouri, where forms resembling the Norway pine are by no means rare. The loblolly, occupying both wet and dry soils, varies accordingly." Cir. No. 12, p. 6.

" . . . It is clear that as all localities have their heavy and their light timber, so they all share in strong and weak, hard and soft material, and the difference in quality of material is evidently far more a matter of individual variation than of soil or climate." *Ibid.*, p. 22

" A representative committee of the Carriage Builders' Association had publicly declared that this important industry could not depend upon the supplies of southern timber, as the oak grown in the South lacked the necessary qualities demanded in carriage construction. Without experiment this statement could be little better than a guess, and was doubly unwarranted, since it condemned an enormous amount of material, and one produced under a great variety of conditions and by at least a dozen species of trees, involving, therefore, a complexity of problems difficult enough for the careful investigator, and entirely beyond the few unsystematic observations of the members of a committee on a flying trip through one of the greatest timber regions of the world.

" A number of samples were at once collected (part of them supplied by the carriage builders' committee), and the fallacy of the broad statement mentioned was fully demonstrated by a short series of tests and a more extensive study into structure and weight of these materials. From these tests it appears that pieces of white oak from Arkansas excelled well-selected pieces from Connecticut, both in stiffness and endwise compression (the two most important forms of resistance)." Report upon the forestry investigations of the U. S. D. A. 1877–1898, p. 331. See also Rep. of Div. of For., 1890, p. 209.

" In some regions there are many small, stunted hickories, which most users will not touch. They have narrow sap, are likely to be birdpecked, and show very slow growth. Yet five of these trees from a steep, dry south slope in West Virginia had an average strength fully equal to that of the pignut from the better situation, and were superior in toughness, the work to maximum load being 36.8 as against 31.2 for pignut. The trees had about twice as many rings per inch as others from better situations.

" This, however, is not very significant, as trees of the same species, age, and size, growing side by side under the same conditions of soil and situation, show great variation in their technical value. It is hard to account for this difference, but it seems that trees growing in wet or moist situations are rather inferior to those growing on fresher soil; also, it is claimed by many hickory users that the wood from limestone soils is superior to that from sandy soils.

" One of the moot questions among hickory men is the relative value of northern and southern hickory. The impression prevails that southern hickory is more porous and brash than hickory from the north. The tests . . . indicate that southern hickory is as tough and strong as northern hickory of the same age. But the southern hickories have a greater tendency to be shaky, and this results in much waste. In trees from southern river bottoms the loss through shakes and grub-holes in many cases amounts to as much as 50 per cent.

" It is clear, therefore, that the difference in northern and southern hickory is not due to geographic location, but rather to the character of timber that is being cut. Nearly all of that from southern river bottoms and from the Cumberland Mountains is from large, old-growth trees; that from the north is from younger trees which are grown under more favorable conditions, and it is due simply to the greater age of the southern trees that hickory from that region is lighter and more brash than that from the north." Bul. 80, pp. 52–55.

SEASON OF CUTTING

It is generally believed that winter-felled timber has decided advantages over that cut at other seasons of the year, and to that cause alone are frequently ascribed much greater durability, less liability to check and split, better color, and even increased strength and toughness. The conclusion from the various experiments made on the subject is that while the time of felling may, and often does, affect the properties of wood, such result is due to the weather conditions rather than to the condition of the wood.

There are two phases of this question. One is concerned with the physiological changes which might take place during the year in the wood of a living tree. The other deals with the purely physical results due to the weather, as differences in temperature, humidity, moisture, and other features to be mentioned later.

Those who adhere to the first view maintain that wood cut in summer is quite different in composition from that cut in winter. One opinion is that in summer the "sap is up," while in winter it is "down," consequently winter-felled timber is drier. A variation of this belief is that in summer the sap contains certain chemicals which affect the properties of wood and does not contain them in winter. Again it is sometimes asserted that wood is actually denser in winter than in summer, as part of the wood substance is dissolved out in the spring and used for plant food, being restored in the fall.

It is obvious that such views could apply only to sapwood, since it alone is in living condition at the time of cutting. Heartwood is dead wood and has almost no function in the existence of the tree other than the purely mechanical one of support. Heartwood does undergo changes, but they are gradual and almost entirely independent of the seasons.

Sapwood might reasonably be expected to respond to seasonal changes, and to some extent it does. Just beneath the bark there is a thin layer of cells which during the growing season have not attained their greatest density. With the exception of this one annual ring, or portion of one, the density of the wood substance of the sapwood is nearly the same the year round. Slight variations may occur due to impregnation with sugar and starch in the winter and its dissolution in the growing season. The time of cutting can have no material effect on the inherent strength and other mechanical properties of wood except in the outermost annual ring of growth.

The popular belief that sap is up in the spring and summer and is down in the winter has not been substantiated by experiment. There are seasonal differences in the composition of sap, but so far as the amount of sap in a tree is concerned there is fully as much, if not more, during the winter than in summer. Winter-cut wood is not drier, to begin with, than summer-felled—in reality, it is likely to be wetter.*

The important consideration in regard to this question is the series of circumstances attending the handling of the timber after it is felled. Wood dries more rapidly in summer than in winter, not because there is less moisture at one time than another, but because of the higher temperature in summer. This greater heat is often accompanied by low humidity, and conditions are favorable for the rapid removal of moisture from the exposed portions of wood. Wood dries by evaporation, and other things being equal, this will proceed much faster in hot weather than in cold.

It is a matter of common observation that when wood dries it shrinks, and if shrinkage is not uniform in all directions the material pulls apart, causing season checks. (See Fig. 27, page 81.) If evaporation proceeds more rapidly on the outside than inside, the greater shrinkage of the outer portions is bound to result in many checks, the number and size increasing with the degree of inequality of drying.

In cold weather, drying proceeds slowly but uniformly, thus allowing the wood elements to adjust themselves with the least amount of rupturing. In summer, drying proceeds rapidly and

---

* See Record, S. J.: Sap in relation to the properties of wood. Proc. Am. Wood Preservers' Assn., Baltimore, Md., 1913, pp. 160–166.

Kempfer, Wm. H.: The air-seasoning of timber. In Bul. 161, Am. Ry. Eng. Assn., 1913, p. 214.

irregularly, so that material seasoned at that time is more likely to split and check.

There is less danger of sap rot when trees are felled in winter because the fungus does not grow in the very cold weather, and the lumber has a chance to season to below the danger point before the fungus gets a chance to attack it. If the logs in each case could be cut into lumber immediately after felling and given exactly the same treatment, for example, kiln-dried, no difference due to the season of cutting would be noted.

WATER CONTENT *

Water occurs in living wood in three conditions, namely: (1) in the cell walls, (2) in the protoplasmic contents of the cells, and (3) as free water in the cell cavities and spaces. In heartwood it occurs only in the first and last forms. Wood that is thoroughly air-dried retains from 8 to 16 per cent of water in the cell walls, and none, or practically none, in the other forms. Even oven-dried wood retains a small percentage of moisture, but for all except chemical purposes, may be considered absolutely dry.

The general effect of the water content upon the wood substance is to render it softer and more pliable. A similar effect of common observation is in the softening action of water on rawhide, paper, or cloth. Within certain limits the greater the water content the greater its softening effect.

Drying produces a decided increase in the strength of wood, particularly in small specimens. An extreme example is the case of a completely dry spruce block two inches in section, which will sustain a permanent load four times as great as that which a green block of the same size will support.

The greatest increase due to drying is in the ultimate crushing strength, and strength at elastic limit in endwise compression; these are followed by the modulus of rupture, and stress at elastic limit in cross-bending, while the modulus of elasticity is least affected. These ratios are shown in Table XV, but it is to be noted

---

* See Tiemann, H. D.: Effect of moisture upon the strength and stiffness of wood. Bul. 70, U. S. Forest Service, Washington, D. C., 1906; also Cir. 108, 1907.

that they apply only to wood in a much drier condition than is used in practice. For air-dry wood the ratios are considerably lower, particularly in the case of the ultimate strength and the elastic limit. Stiffness (within the elastic limit), while following a similar law, is less affected. In the case of shear parallel to the grain, the general effect of drying is to increase the strength, but this is often offset by small splits and checks caused by shrinkage.

TABLE XV

EFFECT OF DRYING ON THE MECHANICAL PROPERTIES OF WOOD, SHOWN IN RATIO OF INCREASE DUE TO REDUCING MOISTURE CONTENT FROM THE GREEN CONDITION TO KILN-DRY (3.5 PER CENT)

(Forest Service Bul. 70, p. 89)

KIND OF STRENGTH	Longleaf pine		Spruce		Chestnut	
	(1)	(2)	(1)	(2)	(1)	(2)
Crushing strength parallel to grain................	2.89	2.60	3.71	3.41	2.83	2.55
Elastic limit in compression parallel to grain...	2.60	2.34	3.80	3.49	2.40	2.26
Modulus of rupture in bending............	2.50	2.20	2.81	2.50	2.09	1.82
Stress at elastic limit in bending............	2.90	2.55	2.90	2.58	2.30	2.00
Crushing strength at right angles to grain........	....	....	2.58	2.48	....	....
Shearing strength parallel to grain............	2.01	1.91	2.03	1.95	1.55	1.47
Modulus of elasticity in compression parallel to grain................	1.63	1.47	2.26	2.08	1.43	1.29
Modulus of elasticity in bending............	1.59	1.35	1.43	1.23	1.44	1.21

NOTE.—The figures in the first column show the relative increase in strength between a green specimen and a kiln-dry specimen of equal size. The figures in the second column show the relative increase of strength of the same block after being dried from a green condition to 3.5 per cent moisture, correction having been made for shrinkage. That is, in the first column the strength values per actual unit of area are used; in the second the values per unit of area of green wood which shrinks to smaller size when dried.

See also Cir. 108, Fig. 1, p. 8.

The moisture content has a decided bearing also upon the manner in which wood fails. In compression tests on very dry specimens the entire piece splits suddenly into pieces before any buckling takes place (see Fig. 9, page 18), while with wet material the block gives way gradually, due to the buckling or bending of

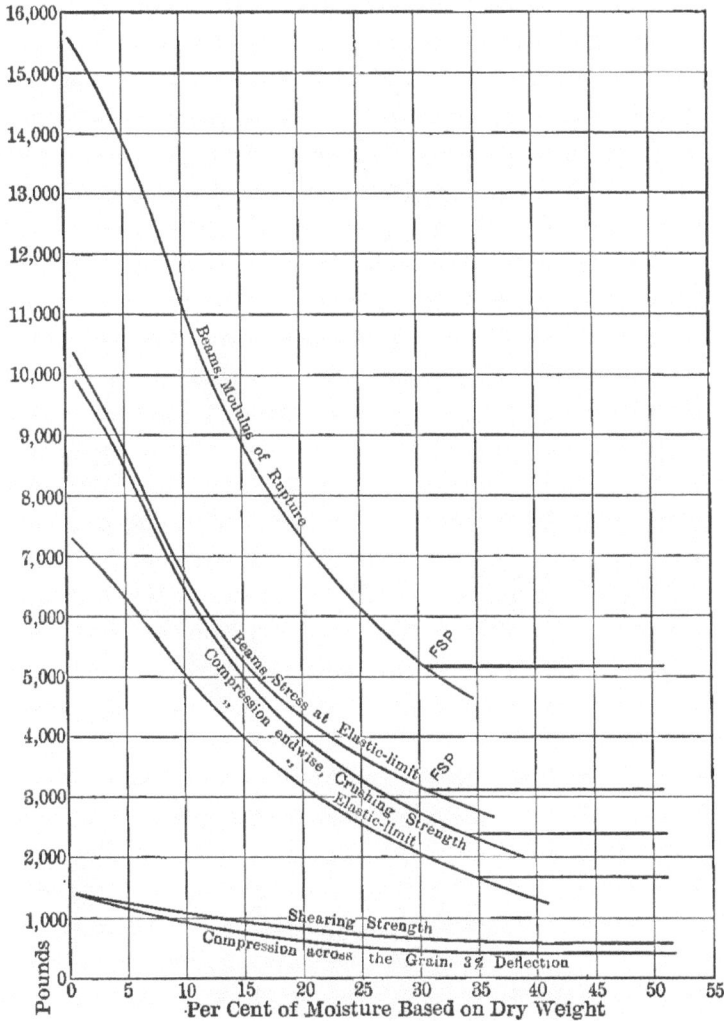

FIG. 24.—Relation of the moisture content to the various strength values of spruce. FSP =fibre-saturation point.

the walls of the fibres along one or more shearing planes. (See Fig. 14, page 21.) In bending tests on wet beams, first failure occurs by compression on top of the beam, gradually extending downward toward the neutral axis. Finally the beam ruptures at the bottom. In the case of very dry beams the failure is usually by splitting or tension on the under side (see Fig. 17, page 35), without compression on the upper, and is often sudden and without warning, and even while the load is still increasing. The effect varies somewhat with different species, chestnut, for example, becoming more brittle upon drying than do ash, hemlock, and longleaf pine. The tensile strength of wood is least affected by drying, as a rule.

In drying wood no increase in strength results until the free water is evaporated and the cell walls begin to dry.* This critical point has been called the *fibre-saturation point.* (See Fig. 24.) Conversely, after the cell walls are saturated with water, any increase in the amount of water absorbed merely fills the cavities and intercellular spaces, and has no effect on the mechanical properties. Hence, soaking green wood does not lessen its strength unless the water is heated, whereupon a decided weakening results.

The strengthening effects of drying, while very marked in the case of small pieces, may be fully offset in structural timbers by inherent weakening effects due to the splitting apart of the wood elements as a result of irregular shrinkage, and in some cases also to the slitting of the cell walls (see Fig. 25). Consequently with large timbers in commercial use it is unsafe to count upon any greater strength, even after seasoning, than that of the green or fresh condition.

In green wood the cells are all intimately joined together and are at their natural or normal size when saturated with water. The cell walls may be considered as made up of little particles with water between them. When wood is dried the films of water between the particles become thinner and thinner until almost entirely gone. As a result the cell walls grow thinner with loss of moisture,—in other words, the cell shrinks.

It is at once evident that if drying does not take place uniformly throughout an entire piece of timber, the shrinkage as a

---

* The wood of *Eucalyptus globulus* (blue gum) appears to be an exception to this rule. Tiemann says: "The wood of blue gum begins to shrink immediately from the green condition, even at 70 to 90 per cent moisture content, instead of from 30 or 25 per cent as in other species of hardwoods." Proc. Soc. Am. For., Washington, Vol. VIII, No. 3, Oct., 1913, p. 313.

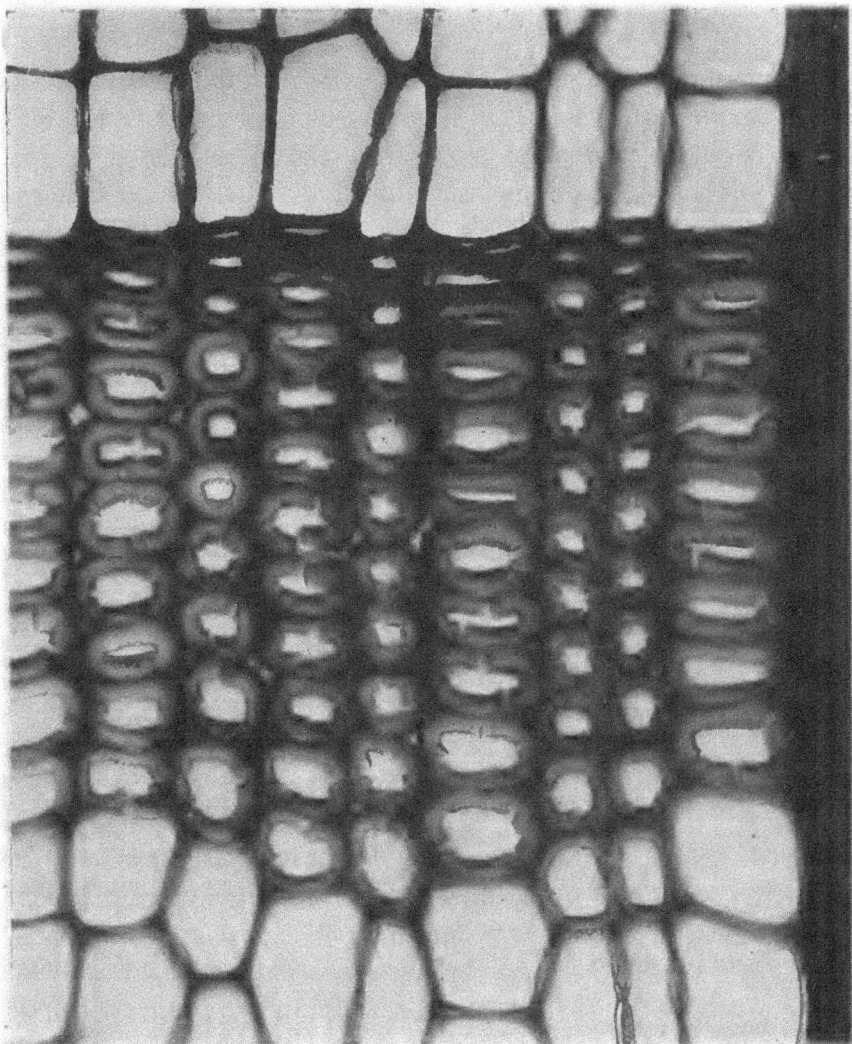

Fig. 25.—Cross section of the wood of western larch showing fissures in the thick-walled cells of the late wood.   Highly magnified.

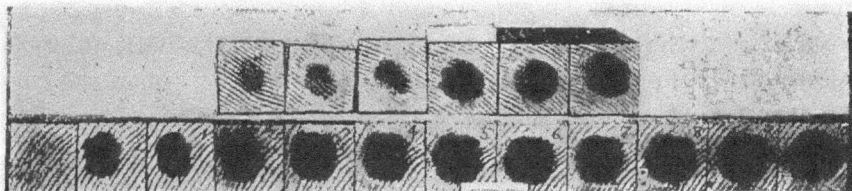

Fig. 26.—Progress of drying throughout the length of a chestnut beam, the black spots indicating the presence of free water in the wood.   The first section at the left was cut one-fourth inch from the end, the next one-half inch, the next one inch, and all the others one inch apart.   The illustration shows case-hardening very clearly.

whole cannot be uniform. The process of drying is from the out-side inward, and if the loss of moisture at the surface is met by a steady capillary current of water from the inside, the shrinkage, so far as the degree of moisture affected it, would be uniform. In the best type of dry kilns this condition is approximated by first heating the wood thoroughly in a moist atmosphere before allowing drying to begin.

In air-seasoning and in ordinary dry kilns this condition too often is not attained, and the result is that a dry shell is formed which encloses a moist interior. (See Fig. 26.) Subsequent drying out of the inner portion is rendered more difficult by this " case-hardened" condition. As the outer part dries it is prevented from shrinking by the wet interior, which is still at its greatest volume. This outer portion must either check open or the fibres become strained in tension. If this outer shell dries while the fibres are thus strained they become " set " in this condition, and are no longer in tension. Later when the inner part dries, it tends to shrink away from the hardened outer shell, so that the inner fibres are now strained in tension and the outer fibres are in compression. If the stress exceeds the cohesion, numerous cracks open up, pro-ducing a " honey-combed " condition, or " hollow-horning," as it is called. If such a case-hardened stick of wood be resawed, the two halves will cup from the internal tension and external com-pression, with the concave surface inward.

For a given surface area the loss of water from wood is always greater from the ends than from the sides, due to the fact that the vessels and other water-carriers are cut across, allowing ready entrance of drying air and outlet for the water vapor. Water does not flow out of boards and timbers of its own accord, but must be evaporated, though it may be forced out of very sappy specimens by heat. In drying a log or pole with the bark on, most of the water must be evaporated through the ends, but in the case of peeled timbers and sawn boards the loss is greatest from the surface because the area exposed is so much greater.

The more rapid drying of the ends causes local shrinkage, and were the material sufficiently plastic the ends would become bluntly tapering. The rigidity of the wood substance prevents this and the fibres are split apart. Later, as the remainder of the stick dries many of the checks will come together, though some of the largest will remain and even increase in size as the drying proceeds. (See Fig. 27.)

A wood cell shrinks very little lengthwise. A dry wood cell is, therefore, practically of the same length as it was in a green or

Photo by U. S. Forest Service.

FIG. 27.—Excessive season checking.

saturated condition, but is smaller in cross section, has thinner walls, and a larger cavity. It is at once evident that this fact makes shrinkage more irregular, for wherever cells cross each other at a decided angle they will tend to pull apart upon drying. This occurs wherever pith rays and wood fibres meet. A considerable portion of every wood is made up of these rays, which for the most part have their cells lying in a radial direction instead of longitudinally. (See Frontispiece.) In pine, over 15,000 of these occur on a square inch of a tangential section, and even in oak the very large rays which are readily visible to the eye as flakes on quarter-sawed material represent scarcely one per cent of the number which the microscope reveals.

A pith ray shrinks in height and width, that is, vertically and tangentially as applied to the position in a standing tree, but very little in length or radially. The other elements of the wood shrink radially and tangentially, but almost none lengthwise or vertically as applied to the tree. Here, then, we find the shrinkage of the rays tending to shorten a stick of wood, while the other cells resist it, and the tendency of a stick to get smaller in circumference is resisted by the endwise reaction or thrust of the rays. Only in a tangential direction, or around the stick in direction of the annual rings of growth, do the two forces coincide. Another factor to the same end is that the denser bands of late wood are continuous in a tangential direction, while radially they are separated by alternate zones of less dense early wood. Consequently the shrinkage along the rings (tangential) is fully twice as much as toward the centre (radial). (See Table XIV, page 56.) This explains why some cracks open more and more as drying advances. (See Fig. 27.)

Although actual shrinkage in length is small, nevertheless the tendency of the rays to shorten a stick produces strains which are responsible for some of the splitting open of ties, posts, and sawed timbers with box heart. At the very centre of a tree the wood is light and weak, while farther out it becomes denser and stronger. Longitudinal shrinkage is accordingly least at the centre and greater toward the outside, tending to become greatest in the sapwood. When a round or a box-heart timber dries fast it splits radially, and as drying continues the cleft widens partly on account of the greater tangential shrinkage and also because the greater contraction of the outer fibres warps the sections apart. If a small hardwood stem is split while green for a short distance at the end and placed where it can dry out rapidly, the sections will become bow-shaped with the concave sides out. These various facts, taken together, explain why, for example, an oak tie, pole, or log may split open its entire length if drying proceeds rapidly and far enough. Initial stresses in the living trees produce a similar effect when the log is sawn into boards. This is especially so in *Eucalyptus globulus* and to a less extent with any rapidly grown wood.

The use of S-shaped thin steel clamps to prevent large checks and splits is now a common practice in this country with cross-ties and poles as it has been for a long time in European countries. These devices are driven into the butts of the timbers so as to cross incipient checks and prevent their widening. In place of the regular S-hook another of crimped iron has been devised. (See Fig. 28.) Thin straps of iron with one tapered edge are run

Photo by U. S. Forest Service.

FIG. 28.—Control of season checking by the use of S-irons.

between intermeshing cogs and crimped, after which they may be cut off any length desired. The time for driving S-irons of either form is when the cracks first appear.

The tendency of logs to split emphasizes the importance of converting them into planks or timbers while in a green condition. Otherwise the presence of large checks may render much lumber worthless which might have been cut out in good condition. The loss would not be so great if logs were perfectly straight-grained, but this is seldom the case, most trees growing more or less spirally or irregularly. Large pieces crack more than smaller ones, quartered lumber less than that sawed through and through, thin pieces, especially veneers, less than thicker boards.

In order to prevent cracks at the ends of boards, small straps of wood may be nailed on them or they may be painted. This method is usually considered too expensive, except in the case of valuable material. Squares used for shuttles, furniture, gunstocks, and tool handles should always be protected at the ends. One of the best means is to dip them into melted paraffine, which seals the ends and prevents loss of moisture there. Another method is to glue paper on the ends. In some cases abroad paper is glued on to all the surfaces of valuable exotic balks. Other substances sometimes employed for the purpose of sealing the wood are grease, carbolineum, wax, clay, petroleum, linseed oil, tar, and soluble glass. In place of solid beams, built-up material is often preferable, as the disastrous results of season checks are thereby largely overcome or minimized.

TEMPERATURE

The effect of temperature on wood depends very largely upon the moisture content of the wood and the surrounding medium. If absolutely dry wood is heated in absolutely dry air the wood expands. The extent of this expanson is denoted by a coefficient corresponding to the increase in length or other dimensions for each degree rise in temperature divided by the original length or other dimension of the specimen. The coefficient of linear expansion of oak has been found to be .00000492; radial expansion, .0000544, or about eleven times the longitudinal. Spruce expands less than oak, the ratio of radial to longitudinal expansion being about six to one. Metals and glass expand equally in all directions, since they are homogeneous substances, while wood is a complicated structure. The coefficient of expansion of iron is .0000285, or nearly six times the coefficient of linear expansion of oak and seven times that of spruce.*

Under ordinary conditions wood contains more or less moisture, so that the application of heat has a drying effect which is accompanied by shrinkage. This shrinkage completely obscures the expansion due to the heating.

Experiments made at the Yale Forest School revealed the effect of temperature on the crushing strength of wet wood. In the case

---

* See Schlich's Manual of Forestry, Vol. V. (rev. ed.), p. 75.

of wet chestnut wood the strength decreases 0.42 per cent for each degree the water is heated above 60° F.; in the case of spruce the decrease is 0.32 per cent.

TABLE XVI

EFFECT OF STEAMING ON THE STRENGTH OF GREEN LOBLOLLY PINE

(Forest Service, Cir. 39)

Treatment	Cylinder conditions			Strength			
	Steaming			Static		Impact	
	Period	Pressure	Temperature	Bending modulus of rupture	Compression parallel to grain	Bending height of drop causing complete failure	Average of the three strengths
	Hrs.	Lbs. per sq. in.	°F.	Per cent	Per cent	Per cent	Per cent
					Untreated wood = 100%		
Steam, at various pressures......	4	...	*230	91.3	79.1	96.4	88.9
	4	10	238	78.2	93.7	93.3	88.4
	4	20	253	83.3	84.2	91.4	86.3
	4	30	269	80.4	78.4	89.8	82.9
	4	40	283	78.1	74.4	74.0	75.5
	4	50	292	75.8	71.5	63.9	70.4
	4	100	337	41.4	65.0	55.2	53.9
Steam, for various periods........	1	20	257	100.6	98.6	86.7	95.3
	2	20	267	88.4	93.0	107.0	96.1
	3	20	260	90.0	93.6	84.1	89.2
	4	20	253	83.3	84.2	91.4	86.3
	5	20	253	85.0	78.1	84.2	82.4
	6	20	242	95.2	89.8	76.0	87.0
	10	20	255	73.7	82.0	76.0	77.2
	20	20	258	67.5	65.0	99.0	77.2

* It will be noted that the temperature was 230°. This is the maximum temperature by the maximum-temperature recording thermometer, and is due to the handling of the exhaust valve. The average temperature was that of exhaust steam.

The effects of high temperature on wet wood are very marked. Boiling produces a condition of great pliability, especially in the case of hardwoods. If wood in this condition is bent and allowed to dry, it rigidly retains the shape of the bend, though its strength may be somewhat reduced. Except in the case of very dry wood the effect of cold is to increase the strength and stiffness of wood. The freezing of any free water in the pores of the wood will augment these conditions.

The effect of steaming upon the strength of cross-ties was investigated by the U. S. Forest Service in 1904. The conclusions were summarized as follows:

" (1) The steam at pressure up to 40 pounds applied for 4 hours, or at a pressure of 20 pounds up to 20 hours, increases the weight of ties. At 40 pounds' pressure applied for 4 hours and at 20 pounds for 5 hours the wood began to be scorched.

" (2) The steamed and saturated wood, when tested immediately after treatment, exhibited weaknesses in proportion to the pressure and duration of steaming. (See Table XVI.) If allowed to air-dry subsequently the specimens regained the greater part of their strength, provided the pressure and duration had not exceeded those cited under (1). Subsequent immersion in water of the steamed wood and dried specimens showed that they were weaker than natural wood similarly dried and resoaked." *

" (3) A high degree of steaming is injurious to wood in strength and spike-holding power. The degree of steaming at which pronounced harm results will depend upon the quality of the wood and its degree of seasoning, and upon the pressure (temperature) of steam and the duration of its application. For loblolly pine the limit of safety is certainly 30 pounds for 4 hours, or 20 pounds for 6 hours." †

Experiments made at the Yale Forest School showed that steaming above 30 pounds' gauge pressure reduces the strength of wood permanently while wet from 25 to 75 per cent.

PRESERVATIVES

The exact effects of chemical impregnation upon the mechanical properties of wood have not been fully determined, though they have been the subject of considerable investigation.‡ More

---

* Cir. 39. Experiments on the strength of treated timber, p. 18.

† *Ibid.*, p. 21. See also Cir. 108, p. 19, table 5.

‡ Hatt, W. K.: Experiments on the strength of treated timber. Cir. 39, U. S. Forest Service, 1906, p. 31.

depends upon the method of treatment than upon the preservatives used. Thus preliminary steaming at too high pressure or for too long a period will materially weaken the wood. (See Temperature, *supra*.)

The presence of zinc chloride does not weaken wood under static loading, although the indications are that the wood becomes brittle under impact. If the solution is too strong it will decompose the wood.

Soaking in creosote oil causes wood to swell, and accordingly decreases the strength to some extent, but not nearly so much so as soaking in water.*

Soaking in kerosene seems to have no significant weakening effect.†

---

* Teesdale, Clyde H.: The absorption of creosote by the cell walls of wood. Cir. 200, U. S. Forest Service, 1912, p. 7.

† Tiemann, H. D.: Effect of moisture upon the strength and stiffness of wood. Bul. 70, U. S. Forest Service, 1907, pp. 122–123, tables 43–44.

PART III

TIMBER TESTING *

PRELIMINARY to making a series of timber tests it is very important that a working plan be prepared as a guide to the investigation. This should embrace: (1) the purpose of the tests; (2) kind, size, condition, and amount of material needed; (3) full description of the system of marking the pieces; (4) details of any special apparatus and methods employed; (5) proposed method of analyzing the data obtained and the nature of the final report. Great care should be taken in the preparation of this plan in order that all problems arising may be anticipated so far as possible and delays and unnecessary work avoided. A comprehensive study of previous investigations along the same or related lines should prove very helpful in outlining the work and preparing the report. (For sample working plan see Appendix, page 127.)

FORMS OF MATERIAL TESTED

In general, four forms of material are tested, namely: (1) large timbers, such as bridge stringers, car sills, large beams, and other pieces five feet or more in length, of actual sizes and grades in common use; (2) built-up structural forms and fastenings, such as built-up beams, trusses, and various kind of joints; (3) small clear pieces, such as are used in compression, shear, cleavage, and small cross-breaking tests; (4) manufactured articles, such as axles, spokes, shafts, wagon-tongues, cross-arms, insulator pins, barrels, and packing boxes.

As the moisture content is of fundamental importance (see Water Content, pages 75–84), all standard tests are usually

---

* The methods of timber testing described here are for the most part those employed by the U. S. Forest Service. See Cir. 38 (rev. ed.), 1909.

made in the green condition. Another series is also usually run in an air-dry condition of about 12 per cent moisture. In all cases the moisture is very carefully determined and stated with the results in the tables.

SIZE OF TEST SPECIMENS

The size of the test specimen must be governed largely by the purpose for which the test is made. If the effect of a single factor, such as moisture, is the object of experiment, it is necessary to use small pieces of wood in order to eliminate so far as possible all disturbing factors. If the specimens are too large, it is impossible to secure enough perfect pieces from one tree to form a series for various tests. Moreover, the drying process with large timbers is very difficult and irregular, and requires a long period of time, besides causing checks and internal stresses which may obscure the results obtained.

On the other hand, the smaller the dimensions of the test specimen the greater becomes the relative effect of the inherent factors affecting the mechanical properties. For example, the effect of a knot of given size is more serious in a small stick than in a large one. Moreover, the smaller the specimen the fewer growth rings it contains, hence there is greater opportunity for variation due to irregularities of grain.

Tests on large timbers are considered necessary to furnish designers data on the probable strength of the different sizes and grades of timber on the market; their coefficients of elasticity under bending (since the stiffness rather than the strength often determines the size of a beam); and the manner of failure, whether in bending fibre stress or horizontal shear. It is believed that this information can only be obtained by direct tests on the different grades of car sills, stringers, and other material in common use.

When small pieces are selected for test they very often are clear and straight-grained, and thus of so much better grade than the large sticks that tests upon them may not yield unit values applicable to the larger sizes. Extensive experiments show, however, (1) that the modulus of elasticity is approximately the same for large timbers as for small clear specimens cut from them, and (2) that the fibre stress at elastic limit for large beams is, except in the weakest timbers, practically equal to the crushing strength of small clear pieces of the same material.*

MOISTURE DETERMINATION

In order for tests to be comparable, it is necessary to know the moisture content of the specimens at the zone of failure. This is determined from disks an inch thick cut from the timber immediately after testing.

In cases, as in large beams, where it is desirable to know not only the average moisture content but also its distribution through the timber, the disks are cut up so as to obtain an outside, a middle, and an inner portion, of approximately equal areas. Thus in a section 10″ x 12″ the outer strip would be one inch wide, and the second one a little more than an inch and a quarter. Moisture determinations are made for each of the three portions separately.

The procedure is as follows:

(1) Immediately after sawing, loose splinters are removed and each section is weighed.

(2) The material is put into a drying oven at 100° C. (212° F.) and dried until the variation in weight for a period of twenty-four hours is less than 0.5 per cent.

(3) The disk is again carefully weighed.

(4) The loss in weight expressed in per cent of the dry weight indicates the moisture content of the specimen from which the specimen was cut.

MACHINE FOR STATIC TESTS

The standard screw machines used for metal tests are also used for wood, but in the case of wood tests the readings must be taken " on the fly," and the machine operated at a uniform speed without interruption from beginning to end of the test. This is on account of the time factor in the strength of wood. (See Speed of Testing Machine, page 92.)

---

* Bul. 108, U. S. Forest Service: Tests of structural timbers, pp. 53–54.

The standard machines for static tests can be used for transverse bending, compression, tension, shear, and cleavage. A common form consists of three main parts, namely: (1) the straining mechanism, (2) the weighing apparatus, and (3) the machinery for communicating motion to the screws.

The straining mechanism consists of two parts, one of which is a movable crosshead operated by four (sometimes two or three) upright steel straining screws which pass through openings in the platform and bear upward on the bed of the machine upon which the weighing platform rests as a fulcrum. At the lower ends of these screws are geared nuts all rotated simultaneously by a system of gears which cause the movable crosshead to rise and fall as desired

The stationary part of the straining mechanism, which is used only for tension and cleavage tests, consists of a steel cage above the movable crosshead and rests directly upon the weighing platform. The top of the cage contains a square hole into which one end of the test specimen may be clamped, the crosshead containing a similar clamp for the other end, in making tension tests.

For testing long beams a special form of machine with an extended platform is used. (See Fig. 29, page 95.)

The weighing platform rests upon knife edges carried by primary levers of the weighing apparatus, the fulcrum being on the bed of the machine, and any pressure upon it is directly transmitted through a series of levers to the weighing beam. This beam is adjusted by means of a poise running on a screw. In operation the beam is kept floating by means of another poise moved back and forth by a screw which is operated by a hand wheel or automatically. The larger units of stress are read from the graduations along the side of the beam, while the intermediate smaller weights are observed on the dial on the rear end of the beam.

The machine is driven by power from a shaft or a motor and is so geared that various speeds are obtainable. One man can operate it.

In making tests the operation of the straining screws is always downward so as to bring pressure to bear upon the weighing platform. For tests in tension and cleavage the specimen is placed between the top of the stationary cage and the movable head and subjected to a pull. For tests in transverse bending, compression, and cleavage the specimen is placed between the movable head and the platform, and a direct compression force applied.

Testing machines are usually calibrated to a portion of their capacity before leaving the factory. The delicacy of the weighing levers is verified by determining the number of pounds necessary to move the beam between the stops while a load of 1,000 pounds rests on the platform. The usual requirement is that ten pounds should accomplish this movement.

The size of machine suitable for compression tests on $2'' \times 2''$ sticks or for $2'' \times 2''$ beams with 26 to 36-inch span has a capacity of 30,000 pounds.

SPEED OF TESTING MACHINE

In instructions for making static tests the rate of application of the stress, i.e., the speed of the machine, is given because the strength of wood varies with the speed at which the fibres are strained. The speed of the crosshead of the testing machine is practically never constant, due to mechanical defects of the apparatus and variations in the speed of the motor, but so long as it does not exceed 25 per cent the results will not be appreciably affected. In fact, a change in speed of 50 per cent will not cause the strength of the wood to vary more than 2 per cent.*

Following are the formulæ used in determining the speed of the movable head of the machine in inches per minute (n):

(1) For endwise compression . . . $n = Z\,l$

(2) For beams (centre loading) . . . $n = \dfrac{Z\,l^2}{6h}$

(3) For beams (third-point loading) . . $n = \dfrac{Z\,l^2}{5.4h}$

Z = rate of fibre strain per inch of fibre length.
l = span of beam or length of compression specimen.
h = height of beam.

---

* See Tiemann, Harry Donald: The effect of the speed of testing upon the strength and the standardization of tests for speed. Proc. Am. Soc. for Testing Materials, Vol. VIII, Philadelphia, 1908.

TABLE XVII

SPEED-STRENGTH MODULI AND RELATIVE INCREASE IN STRENGTH AT RATES OF FIBRE STRAIN INCREASING IN GEOMETRICAL RATIO. (Tiemann, *loc. cit.*)

(Values in parentheses are approximate)

	2/3			2			6			18			54			162			486		
Rate of fibre strain. Ten-thousandths in. per min. per in.	2/3			2			6			18			54			162			486		
COMPRESSION — Speed of cross-head. Ins. per min.	0.000383			0.00115			0.00345			0.0103			0.0310			0.0931			0.279		
Specimens	Wet	Dry	All	Wet	Dry	All	Wet	Dry	All	Wet	Dry	All	Wet	Dry	All	Wet	Dry	All	Wet	Dry	All
Relative crushing strength	…	…	…	100.0	100.0	100.0	103.4	100.8	101.5	107.5	102.7	103.8	113.9	105.5	107.9	121.3	108.3	116.4	128.8	110.0	118.9
Speed-strength modulus, T	…	…	…	0.017	(0.006)	(0.009)	0.033	0.012	0.016	0.047	0.021	0.029	0.053	0.027	0.039	0.060	0.023	0.049	(0.052)	(0.015)	(0.040)
BENDING — Speed of cross-head. Ins. per min.	0.0072			0.0216			0.0648			0.194			0.583			1.75			5.25		
Specimens	Wet	Dry	All	Wet	Dry	All	Wet	Dry	All	Wet	Dry	All	Wet	Dry	All	Wet	Dry	All	Wet	Dry	All
Relative modulus of rupture	97.4	99.0	98.2	100.0	100.0	100.0	105.1	102.1	103.7	111.3	105.8	108.1	117.9	108.6	112.7	123.7	109.6	116.3	126.3	110.3	118.9
Speed-strength modulus, T	(0.014)	(0.005)	0.012	0.033	0.014	0.026	0.049	0.026	0.037	0.053	0.033	0.038	0.049	0.014	0.035	0.038	0.006	0.025	0.023	0.004	0.014

NOTE.—The usual speeds of testing at the U. S. Forest Service laboratory are at rates of fibre strain of 15 and 10 ten-thousandths in. per min. per in. for compression and bending respectively.

The values commonly used for Z are as follows:

Bending large beams . . . . $Z = 0.0007$
Bending small beams . . . . $Z = 0.0015$
Endwise compression—large specimens $Z = 0.0015$
Endwise compression—small   " $Z = 0.003$
Right-angled compression—large " $Z = 0.007$
Right-angled compression—small " $Z = 0.015$
Shearing parallel to the grain . . $Z = 0.015$

Example: At what speed should the crosshead move to give the required rate of fibre strain in testing a small beam $2'' \times 2'' \times 30''$. (Span $= 28''$.) Substituting these values in equation (2) above:

$$n = \frac{0.0015 \times 28^2}{6 \times 2} = 0.1 \text{ inch per minute.}$$

In order that tests may be intelligently compared, it is important that account be taken of the speed at which the stress was applied. In determining the basis for a ratio between time and strength the rate of strain, which is controllable, and not the ratio of stress, which is circumstantial, should be used. In other words, the rate at which the movable head of the testing machine descends and not the rate of increase in the load is to be regulated. This ratio, to which the name *speed-strength modulus* has been given, may be expressed as a coefficient which, if multiplied into any proportional change in speed, will give the proportional change in strength. This ratio is derived from empirical curves. (See Table XVII.)

BENDING LARGE BEAMS

Apparatus: A static bending machine (described above), with a special crosshead for third-point loading and a long platform bearing knife-edge supports, is required. (See Fig. 29.)

Preparing the material: Standard sizes and grades of beams and timbers in common use are employed. The ends are roughly squared and the specimen weighed and measured, taking the cross-sectional dimensions midway of the length. Weights should be to the nearest pound, lengths to the nearest 0.1 inch, and cross-sectional dimensions to the nearest 0.01 inch.

Marking and sketching: The butt end of the beam is marked

A and the top end B. While facing A, the top side is marked a, the right hand b, the bottom c, the left hand d. Sketches are made of each side and end, showing (1) size, location, and condition of knots, checks, splits, and other defects; (2) irregularities of grain; (3) distribution of heartwood and sapwood; and on the ends: (4) the location of the pith and the arrangement of the growth rings, (5) number of rings per inch, and (6) the proportion of late wood.

Photo by U. S. Forest Service.

FIG. 29.—Static bending test on large beam. Note arrangement of wire and scale for measuring deflection; also method of applying load at "third-points."

The number of rings per inch and the proportion of late wood should always be determined along a radius or a line normal to the rings. The average number of rings per inch is the total number of rings divided by the length of the line crossing them. The proportion of late wood is equal to the sum of the widths of the late wood crossed by the line, divided by the length of the line. Rings per inch should be to the nearest 0.1; late wood to the nearest 0.1 per cent.

Since in large beams a great variation in rate of growth and

relative amount of late wood is likely in different parts of the
section, it is advisable to consider the cross section in three
volumes, namely, the upper and lower quarters and the middle
half. The determination should be made upon each volume separately, and the average for the entire cross section obtained
from these results.

At the conclusion of the test the failure, as it appears on each
surface, is traced on the sketches, with the failures numbered in
the order of their occurrence. If the beam is subsequently cut
up and used for other tests an additional sketch may be desirable
to show the location of each piece.

Adjusting specimen in machine: The beam is placed in the
machine with the side marked *a* on top, and with the ends projecting equally beyond the supports. In order to prevent crushing
of the fibre at the points where the stress is applied it is necessary
to use bearing blocks of maple or other hard wood with a convex
surface in contact with the beam. Roller bearings should be
placed between the bearing blocks and the knife edges of the
crosshead to allow for the shortening due to flexure. (See Fig. 29.)
Third-point loading is used, that is, the load is applied at two points
one-third the span of the beam apart. (See Fig. 30.) This affords
a uniform bending moment throughout the central third of the
beam.

Measuring the deflection: The method of measuring the deflection should be such that any compression at the points of support
or at the application of the load will not affect the reading. This
may be accomplished by driving a small nail near each end of
the beam, the exact location being on the neutral plane and vertically above each knife-edge support. Between these nails a fine
wire is stretched free of the beam and kept taut by means of a
rubber band or coiled spring on one end. Behind the wire at a
point on the beam midway between the supports a steel scale
graduated to hundredths of an inch is fastened vertically by
means of thumb-tacks or small screws passing through holes in it.
Attachment should be made on the neutral plane.

The first reading is made when the scale beam is balanced at
zero load, and afterward at regular increments of the load which
is applied continuously and at a uniform speed. (See Speed of

Testing Machine, page 92.) If desired, however, the load may be read at regular increments of deflection. The deflection readings should be to the nearest 0.01 inch. To avoid error due to parallax, the readings may be taken by means of a reading telescope about ten feet distant and approximately on a level with the wire. A mirror fastened to the scale will increase the accuracy of the readings if the telescope is not used. As in all tests on timber, the

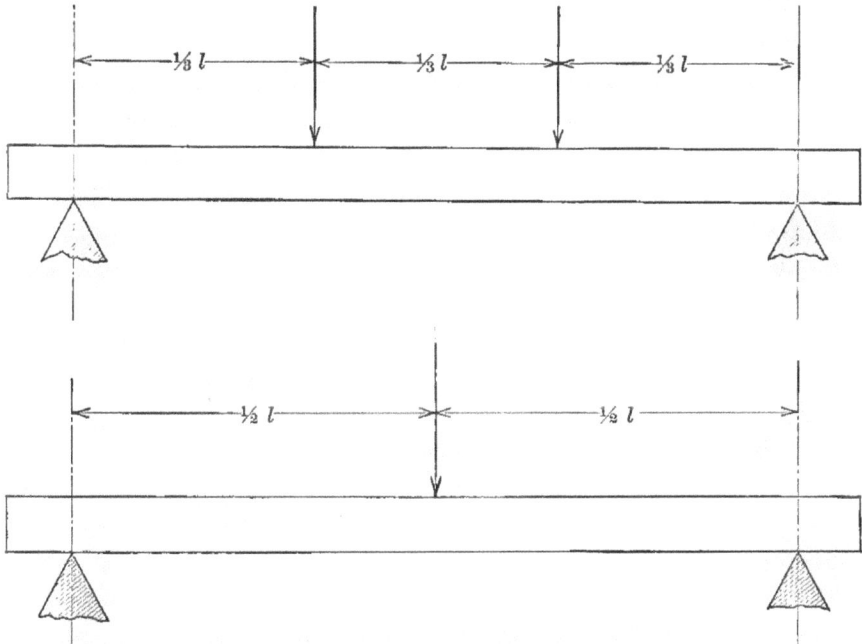

FIG. 30.—Two methods of loading a beam, namely, third-point loading (upper), and centre loading (lower).

strain must be continuous to rupture, not intermittent, and readings must be taken " on the fly." The weighing beam is kept balanced after the yield point is reached and the maximum load, and at least one point beyond it, noted.

Log of the test: The proper log sheet for this test consists of a piece of cross-section paper with space at the margin for notes. (See Fig. 32, page 101.) The load in some convenient unit (1,000 to 10,000 pounds, depending upon the dimensions of the specimen) is entered on the ordinates, the deflection in tenths of an inch on the abscissæ. The increments of load should be chosen so as to furnish about ten points on the stress-strain diagram below the elastic limit.

As the readings of the wire on the scale are made they are entered directly in their proper place on the cross-section paper. In many cases a test should be continued until complete failure results. The points where the various failures occur are indicated on the stress-strain diagram. A brief description of the failure is made on the margin of the log sheet, and the form traced on the sketches.

Disposal of the specimen: Two one-inch sections are cut from the region of failure to be used in determining the moisture content. (See Moisture Determination, page 90.) A two-inch section may be cut for subsequent reference and identification, and possible microscopic study. The remainder of the beam may be cut into small beams and compression pieces.

Calculating the results: The formulæ used in calculating the results of tests on large rectangular simple beams loaded at third points of the span are as follows:

$$(1) \quad J = \frac{0.75\,P}{b\,h} \qquad\qquad (3) \quad R = \frac{l\,(P + 0.75\,W)}{b\,h^2}$$

$$(2) \quad r = \frac{l\,(P_1 + 0.75\,W)}{b\,h^2} \qquad (4) \quad E = \frac{P_1\,l^3}{4.7\,D\,b\,h^3}$$

$$(5) \quad S = \frac{0.87\,P_1\,D}{2\,V}$$

b, h, l = breadth, height, and span of specimen, inches.

D = total deflection at elastic limit, inches.

P = maximum load, pounds.

P_1 = load at elastic limit, pounds.

E = modulus of elasticity, pounds per square inch.

r = fibre stress at elastic limit, pounds per sq. inch.

R = modulus of rupture, pounds per square inch.

S = elastic resilience or work to elastic limit, inch-pounds per cu. in.

J = greatest calculated longitudinal shear, pounds per square inch.

V = volume of beam, cubic inches.

W = weight of the beam.

In large beams the weight should be taken into account in calculating the fibre stress. In (2) and (3) three-fourths of the weight of the beam is added to the load for this reason.

BENDING SMALL BEAMS

Apparatus: An ordinary static bending machine, a steel I-beam bearing two adjustable knife-edge supports to rest on the platform, and a special deflectometer, are required. (See Fig. 31.)

Preparing the material: The specimens may be of any convenient size, though beams $2'' \times 2'' \times 30''$ tested over a 28-inch span,

Photo by U. S. Forest Service.

FIG. 31.—Static bending test on small beam. Note the use of the deflectometer with indicator and dial for measuring the deflection; also roller bearings between beam and supports.

are considered best. The beams are surfaced on all four sides, care being taken that they are not damaged by the rollers of the surfacing machine. Material for these tests is sometimes cut from large beams after failure. The specimens are carefully weighed in grams, and all dimensions measured to the nearest 0.01 inch. If to be tested in a green or fresh condition the specimens should be kept in a damp box or covered with moist sawdust until needed. No defects should be allowed in these specimens.

Marking and sketching: Sketches are made of each end of the specimen to show the character of the growth, and after testing, the manner of failure is shown for all four sides. In obtaining data regarding the rate of growth and the proportion of late wood the same procedure is followed as with large beams.

Adjusting specimen in machine: The beam should be correctly centred in the machine and each end should have a plate with roller bearings between it and the support. Centre loading is used. Between the movable head of the machine and the specimen is placed a bearing block of maple or other hard wood, the lower surface of which is curved in a direction along the beam, the curvature of which should be slightly less than that of the beam at rupture, in order to prevent the edges from crushing into the fibres of the test piece.

Measuring the deflection: The method of measuring deflection of large beams can be used for small sizes, but because of the shortness of the span and consequent slight deformation in the latter, it is hardly accurate enough for good work. The special deflectometer shown in Fig. 31 allows closer reading, as it magnifies the deflection ten times. It rests on two small nails driven in the beam on the neutral plane and vertically above the supports. The fine wire on the wheel at the base of the indicator is attached to another small nail driven in the beam on the neutral plane midway between the end nails. All three nails should be in place before the beam is put into the machine. The indicator is adjustable by means of a thumb-screw at the base and is set at zero before the load is applied. Deflections are read to the nearest 0.001 inch. For rate of application of load see Speed of Testing Machine, page 92. The speed should be uniform from start to finish without stopping. Readings must be made " on the fly."

Log of the test: The log sheets used for small beams (see Fig. 32) are the same as for large sizes and the procedure is practically identical. The stress-strain diagram is continued to or beyond the maximum load, and in a portion of the tests should be continued to six-inch deflection or until the specimen fails to support a load of 200 pounds. Deflection readings for equal increments of

FORM 512
(Supersedes Form 176)

Timber Test Log Sheet

Project No. ___ 124 _____

U.S. DEPARTMENT OF AGRICULTURE
FOREST SERVICE

Working Plan No. __ 124 __

Ship. No. ___ 314 __   Stick No. __ N 7 __

Station   Madison   Date   3/10/14

Laboratory No. _____

Piece No. ___ 34 ____   Mark ___ b _____

Species ___ W. Yellow Pine _____

Kind of test ___ Bending _____

Grade ____ Clear _____

Group _____

Loading ____ Center _____

Span _____ 28˚ _____

Distance between collars _____

Width of plate _____

Machine ____ M-103 _____

Speed of mach. _ 0.105_ in. per min. _____

Weight of hammer _____

Height ___ 2.01˝ _____

Width ___ 2.01˝ _____

Length ___ 29.83˝ _____

Cross section _____

Weight ___ 953 grams _____

Rings per inch ___ 21 _____

Sap _____

Summerwood _____

Seasoning __ Air dry _____

Moisture ___ 9.3% _____

Kind of failure __ Compression ___

_____ followed by _____

_____ simple tension _____

Remarks _____

_____

_____

Sketch

8—580

Max. load   3330 lbs.
Def. at max. load   .53 in.
Load at E. L.   1900 lbs.
Def. at E. L.   0.305 in.
Max. drop

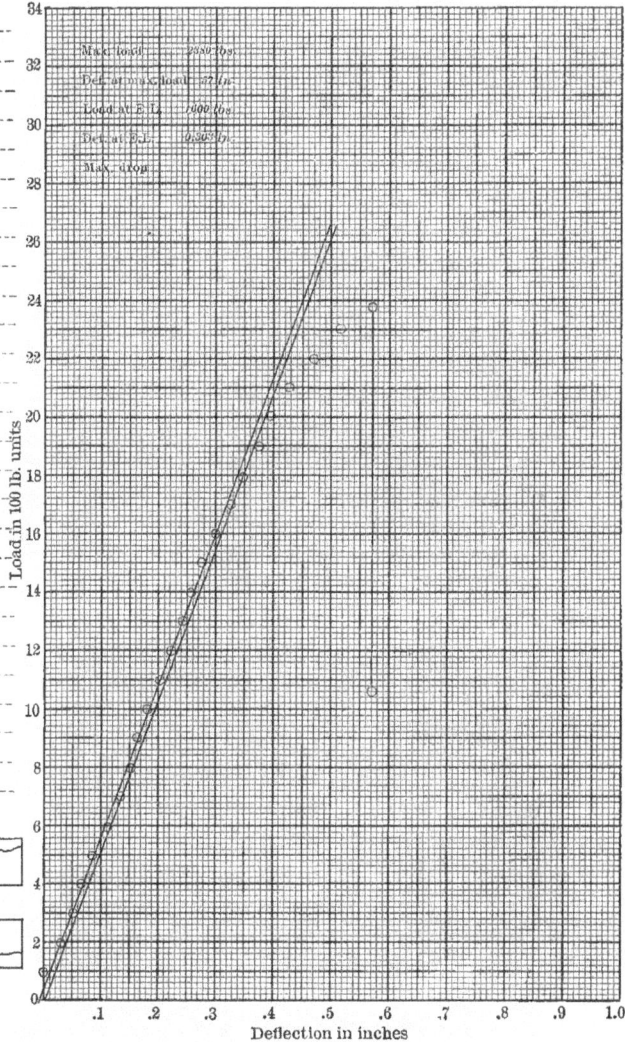

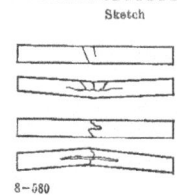

FIG. 32.—Sample log sheet, giving full details of a transverse bending test on a small pine beam.

load are taken until well beyond the elastic limit, after which the scale beam is kept balanced and the load read for each 0.1 inch deflection. The load and deflection at first failure, the maximum load, and any points of sudden change should be shown on the diagram, even though they do not occur at one of the regular points. A brief description of the failure and the nature of any defects is entered on the log sheet.

Calculating the results: The formulæ used in calculating the results of tests on small rectangular simple beams are as follows:

$$(1) \quad J = \frac{0.75\,P}{b\,h} \qquad\qquad (3) \quad R = \frac{1.5\,P\,l}{b\,h^2}$$

$$(2) \quad r = \frac{1.5\,P_1\,l}{b\,h^2} \qquad\qquad (4) \quad E = \frac{P_1\,l^3}{4\,D\,b\,h^3}$$

$$(5) \quad S = \frac{P_1\,D}{2\,V}$$

The same legend is used as on page 98. The weight of the beam itself is disregarded.

ENDWISE COMPRESSION

Apparatus: An ordinary static testing machine and a compressometer are required. (See Fig. 33.)

Preparing the material: Two classes of specimens are commonly used, namely, (1) posts 24 inches in length, and (2) small clear blocks approximately $2'' \times 2'' \times 8''$. The specimens are surfaced on all four sides and both ends squared smoothly and evenly. They are carefully weighed, measured, rate of growth and proportion of late wood determined, as in bending tests. After the test a moisture section is cut and weighed. Ordinarily these specimens should be free from defects.

Sketching: Sketches are made of each end of the specimens to show the character of the growth. After testing, the manner of failure is shown for all four sides, and the various parts of the failure are numbered in the order of their occurrence.

Adjusting specimen in machine: The compressometer collars are adjusted, the distance between them being 20 inches for the posts and 6 inches for the blocks. If the two ends of the blocks are not exactly parallel a ball-and-socket block can be placed between the upper end of the specimen and the movable head of the machine to overcome the irregularity. If the blocks are true they can simply be stood on end upon the platform and the movable head allowed to press directly upon the upper end.

Measuring the deformation: The deformation is measured by a compressometer. (See Fig. 33.) The latter registers to 0.001 inch. In the case of posts the compression between the collars is com-

Photo by *U. S. Forest Service.*

Fig. 33.—Endwise compression test, showing method of measuring the deformation by means of a compressometer.

municated to the four points on the arms by means of brass rods; with short blocks, as in Fig. 33, the points of the arms are in direct contact with the collars. The operator lowers the fulcrum of the apparatus by moving the micrometer screws at such a rate that the set-screw in the rear end of the upper lever is kept barely touching the fixed arm below it, being guided by a bell operated by electric contact.

Log of the test: The load is applied continuously at a uniform rate of speed. (See Speed of Testing Machine, page 92.)

Readings are taken from the scale of the compressometer at regular increments of either load or compression. The stress-strain diagram is continued to at least one deformation point beyond the maximum load, and in event of sudden failure, the direction of the curve beyond the maximum point is indicated. A brief description of the failure is entered on the log sheet. (See Fig. 34.)

In short specimens the failure usually occurs in one or several planes diagonal to the axis of the specimen. If the ends are more moist than the middle a crushing may occur on the extreme ends in a horizontal plane. Such a test is not valid and should always be culled. If the grain is diagonal or the stress is unevenly applied a diagonal shear may occur from top to bottom of the test specimen. Such tests are also invalid and should be culled. When the plane (or several planes) of failure occurs through the body of the specimen the test is valid. It may sometimes be advantageous to allow the extreme ends to dry slightly before testing in order to bring the planes of failure within the body. This is a perfectly legitimate procedure provided no drying is allowed from the sides of the specimen, and the moisture disk is cut from the region of failure.

Calculating the results: The formulæ used in calculating the results of tests on endwise compression are as follows:

$$(1) \quad C = \frac{P}{A} \qquad\qquad (3) \quad E = \frac{P_1\, l}{A\, D}$$

$$(2) \quad c = \frac{P_1}{A} \qquad\qquad (4) \quad S = \frac{P\, D}{2\, V}$$

C = crushing strength, pounds per square inch.
c = fibre strength at elastic limit, pounds per square inch.
A = area of cross section, square inches.
l = distance between centres of collars, inches.
D = total shortening at elastic limit, inches.
V = volume of specimen, cubic inches.
Remainder of legend as on page 98.

COMPRESSION ACROSS THE GRAIN

Apparatus: An ordinary static testing machine, a bearing plate, and a deflectometer are required. (See Fig. 35.)

FORM 512
(Supersedes Form 176)

Project No._124_

Working Plan No._124_

Laboratory No._ _ _ _ _ _

Species_ W. Yellow Pine _ _ _ _ _ _

Kind of test_ Comp. // to gr._ _ _ _ _ _

Grade_ _ _ _ Clear _ _ _ _ _ _ _ _ _ _

Group_ _ _ _ _ _ _ _ _ _ _ _ _ _ _ _ _

Loading _ _ _ _ _ _ _ _ _ _ _ _ _ _ _ _

Span_ _ _ _ _ _ _ _ _ _ _ _ _ _ _ _ _ _

Distance between collars_ _ _ _ 6" _ _ _ _ _

Width of plate_ _ _ _ _ _ _ _ _ _ _ _ _ _

Machine_ _ M-1040 _ _ _ _ _ _ _ _ _ _

Speed of mach._ 0.024 _ in. per min._ _ _ _

Weight of hammer_ _ _ _ _ _ _ _ _ _ _ _ _

Height_ _ _ _ _ _ _ _ _ _ _ _ _ _ _ _

Width_ _ _ _ _ _ _ _ _ _ _ _ _ _ _ _

Length_ _ 8.00" _ _ _ _ _ _ _ _ _ _ _

Cross-section_ 2.00" x 2.00" _ _ _ _ _ _

Weight_ 216 grams _ _ _ _ _ _ _ _ _ _

Rings per inch_ 14 _ _ _ _ _ _ _ _ _ _

Sap _ _ _ _ _ _ _ _ _ _ _ _ _ _ _ _ _

Summerwood_ _ _ _ _ _ _ _ _ _ _ _ _ _

Seasoning_ _ Air dry _ _ _ _ _ _ _ _ _

Moisture_ _ _ _ 7.8% _ _ _ _ _ _ _ _

Kind of failure _ _ Crushing _ _ _ _ _ _

_ _ _ _ _ _ at the bottom _ _ _ _ _ _ _

_ _ _ _ _ _ _ _ _ _ _ _ _ _ _ _ _ _ _ _

Remarks _ _ _ _ _ _ _ _ _ _ _ _ _ _ _

_ _ _ _ _ _ _ _ _ _ _ _ _ _ _ _ _ _ _ _

Sketch

8-580

U.S. DEPARTMENT OF AGRICULTURE
FOREST SERVICE

Station Madison   Date 3/10/14

Timber Test Log Sheet

Ship. No._244_ _ _   Stick No._N7_ _ _ _

Piece No._34_ _ _ _   Mark_ _b1_ _ _ _

Max. load 48,000 lbs.

Def. at max. load

Load at E. L. 22,000 lbs.

Def. at E. L. 0.0208 in.

Max. drop

Load in 1000 lb. units

Compression in inches

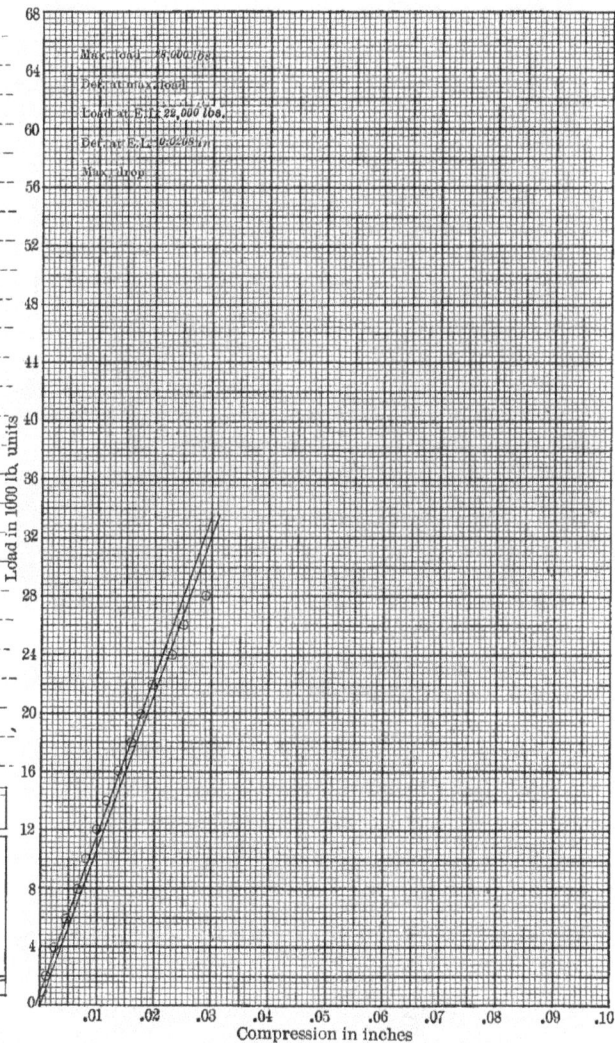

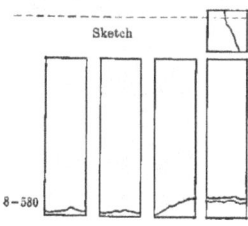

FIG. 34.—Sample log sheet of an endwise compression test on a short pine column.

Preparing the material: Two classes of specimens are used, namely, (1) sections of commercial sizes of ties, beams, and other timbers, and (2) small, clear specimens with the length several times the width. Sometimes small cubes are tested, but the results are hardly applicable to conditions in practice. In (2) the sides are surfaced and the ends squared. The specimens are then carefully measured and weighed, defects noted, rate of growth

FIG. 35.—Compression across the grain. Note method of measuring the deformation by means of a deflectometer.

and proportion of late wood determined, as in bending tests. (See page 95.)   After the test a moisture section is cut and weighed.

Sketching : Sketches are made as in endwise compression tests.   (See page 102.)

Adjusting specimen in machine: The specimen is laid horizontally upon the platform of the machine and a steel bearing plate placed on its upper surface immediately beneath the centre of the movable head. For the larger specimens this plate is six inches wide; for the smaller sizes, two inches wide. The plate in all cases projects over the edges of the test piece, and in no case should the length of the latter be less than four times the width of the plate.

Measuring the deformation: The compression is measured by means of a deflectometer (see Fig. 35), which, after the first increment of load is applied, is adjusted (by means of a small set screw) to read zero. The actual downward motion of the movable head (corresponding to the compression of the specimen) is multiplied ten times on the scale from which the readings are made.

Log of the test: The load is applied continuously and at uniform speed (see Speed of Testing Machine, p. 92), until well beyond the elastic limit. The compression readings are taken at regular load increments and entered on the cross-section paper in the usual way. Usually there is no real maximum load in this case, as the strength continually increases as the fibres are crushed more compactly together.

Calculating the results: Ordinarily only the fibre stress at the elastic limit (c) is computed. It is equal to the load at elastic limit (P_1) divided by the area under the plate (B). $\left(c = \dfrac{P_1}{B} \right)$

SHEAR ALONG THE GRAIN

Apparatus: An ordinary static testing machine and a special tool designed for producing single shear are required. (See Figs. 36 and 37.) This shearing apparatus consists of a solid steel frame with set screws for clamping the block within it firmly in a vertical position. In the centre of the frame is a vertical slot in which a square-

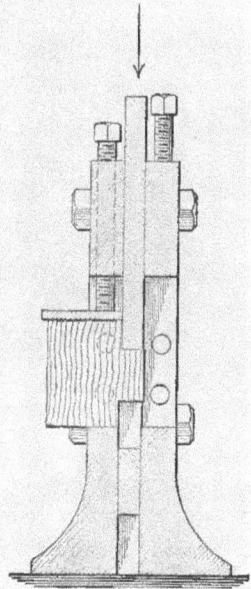

FIG. 36.—Vertical section of shearing tool.

edged steel plate slides freely. When the testing block is in position, this plate impinges squarely along the upper surface of the tenon or lip, which, as vertical pressure is applied, shears off.

Preparing the material: The specimens are usually in the form of small, clear, straight-grained blocks with a projecting tenon or lip to be sheared off. Two common forms and sizes are shown in Figure 38. Part of the blocks are cut so that the shearing surface is parallel to the growth rings, or tangential; others at

FIG. 37.—Front view of shearing tool with test specimen and steel plate in position for testing.

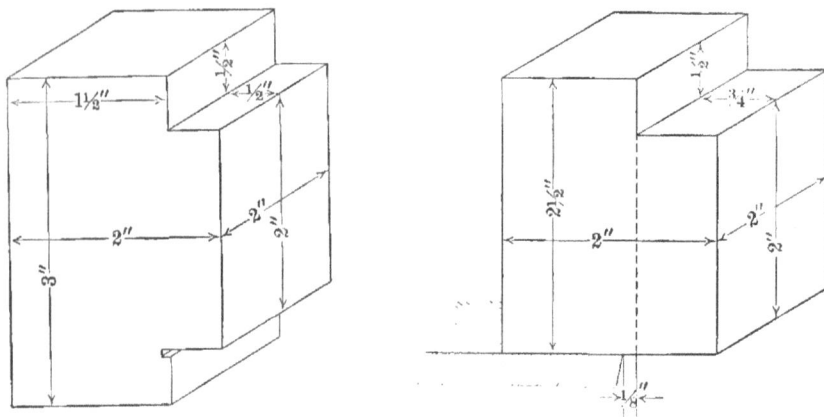

FIG. 38.—Two forms of shear test specimens.

right angles to the growth rings, or radial. It is important that the upper surface of the tenon or lip be sawed exactly parallel to the base of the block. When the form with a tenon is used the under cut is extended a short distance horizontally into the block to prevent any compression from below.

In designing a shearing specimen it is necessary to take into consideration the proportions of the area of shear, since, if the

Photo by U. S. Forest Service.
FIG. 39.—Making a shearing test.

length of the portion to be sheared off is too great in the direction of the shearing face, failure would occur by compression before the piece would shear. Inasmuch as the endwise compressive strength is sometimes not more than five times the shearing strength, the shearing surface should be less than five times the surface to which the load is applied. This condition is fulfilled in the specimens illustrated.

Shearing specimens are frequently cut from beams after testing. In this case the specific gravity (dry), proportion of late wood, and rate of growth are assumed to be the same as already recorded for the beams. In specimens not so taken, these quantities are determined in the usual way. The sheared-off portion is used for a moisture section.

Adjusting specimen in machine: The test specimen is placed in the shearing apparatus with the tenon or lip under the sliding plate, which is centred under the movable head of the machine. (See Fig. 39.) In order to reduce to a minimum the friction due to the lateral pressure of the plate against the bearings of the slot, the apparatus is sometimes placed upon several parallel steel rods to form a roller base. A slight initial load is applied to take up the lost motion of the machinery, and the beam balanced.

Log of the test: The load is applied continuously and at a uniform rate until failure, but no deformations are measured. The points noted are the maximum load and the length of time required to reach it. Sketches are made of the failure. If the failure is not pure shear the test is culled.

The shearing strength per square inch is found by dividing the maximum load by the cross-sectional area. $\left(Q = \dfrac{P}{A}\right)$

IMPACT TEST

Apparatus: There are several types of impact testing machines.* One of the simplest and most efficient for use with wood is illustrated in Figure 40. The base of the machine is 7 feet long, 2.5 feet wide at the centre, and weighs 3,500 pounds. Two upright columns, each 8 feet long, act as guides for the striking head. At the top of the column is the hoisting mechanism for raising or lowering the striking weights. The power for operating the machine is furnished by a motor set on the top. The hoisting mechanism is all controlled by a single operating lever, shown on the side of the column, whereby the striking weight may be raised, lowered, or stopped at the will of the operator. There is an automatic safety device for stopping the machine when the weight reaches the top.

The weight is lifted by a chain, one end of which passes over a

* For description of U. S. Forest Service automatic and autographic impact testing machine, see Proc. Am. Soc. for Testing Materials, Vol. VIII, 1908, pp. 538–540.

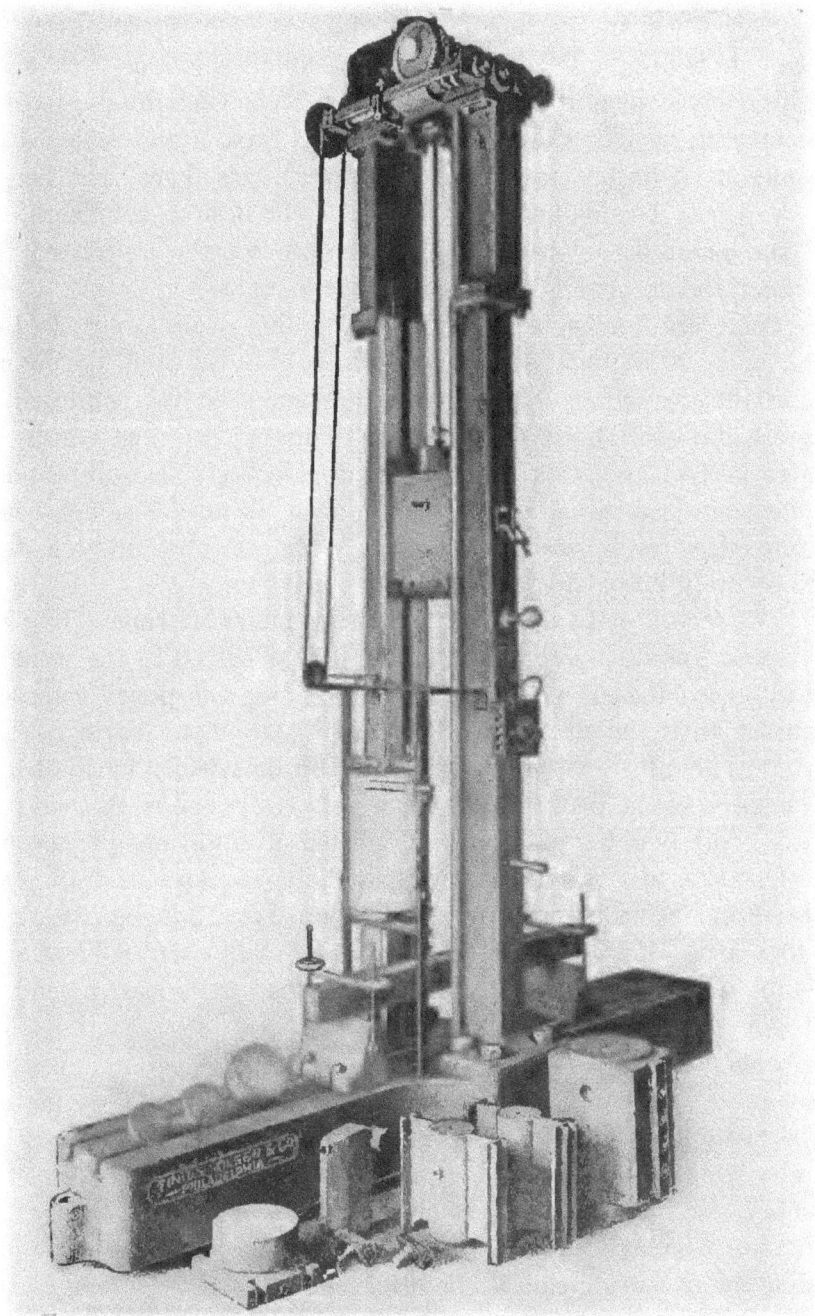

FIG. 40.—Impact testing machine.

sprocket wheel in the hoisting mechanism. On the lower end of the chain is hung an electro-magnet of sufficient magnetic strength to support the heaviest striking weights. When it is desired to drop the striking weight the electric current is broken and reversed by means of an automatic switch and current breaker. The height of drop may be regulated by setting at the desired height on one of the columns a tripping pin which throws the switch on the magnet and so breaks and reverses the current.

There are four striking weights, weighing respectively 50, 100, 250, and 500 pounds, any one of which may be used, depending upon the desired energy of blow. When used for compression tests a flat steel head six inches in diameter is screwed into the lower end of the weight. For transverse tests, a well-rounded knife edge is screwed into the weight in place of the flat head. Knife edges for supporting the ends of the specimen to be tested, are securely bolted to the base of the machine.

The record of the behavior of the specimen at time of impact is traced upon a revolving drum by a pencil fixed in the striking head. (See Fig. 41.) When a drop is made the pencil comes in contact with the drum and is held in place by a spring. The drum is revolved very slowly, either automatically or by hand. The speed of the drum can be recorded by a pencil in the end of a tuning fork which gives a known number of vibrations per second.

One size of this machine will handle specimens for transverse tests 9 inches wide and 6-foot span; the other, 12 inches wide and 8-foot span. For compression tests a free fall of about 6.5 feet may be obtained. For transverse tests the fall is a little less, depending upon the size of the specimen.

The machine is calibrated by dropping the hammer upon a copper cylinder. The axial compression of the plug is noted. The energy used in static tests to produce this axial compression under stress in a like piece of metal is determined. The external energy of the blow (*i.e.*, the weight of the hammer × the height of drop) is compared with the energy used in static tests at equal amounts of compression. For instance:

Energy delivered, impact test . . 35,000 inch-pounds
Energy computed from static test . 26,400 " "
Efficiency of blow of hammer . . 75.3 per cent.

Preparing the material: The material used in making impact tests is of the same size and prepared in the same way as for static bending and compression tests. Bending in impact tests is more commonly used than compression, and small beams with 28-inch span are usually employed.

Method: In making an impact bending test the hammer is

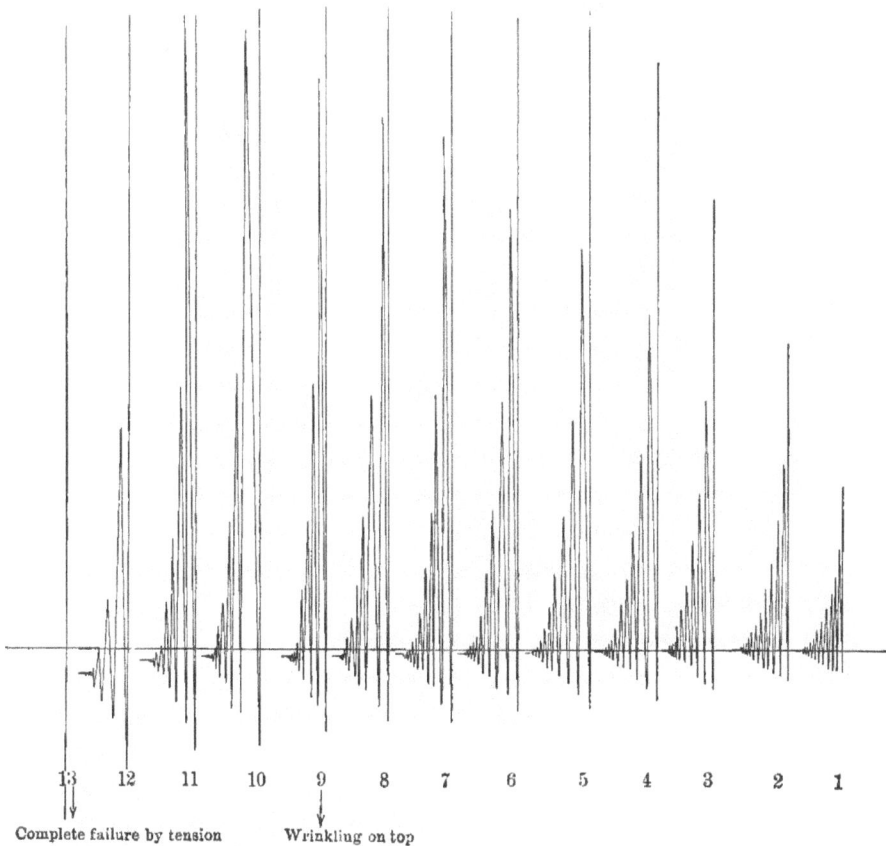

Complete failure by tension   Wrinkling on top

FIG. 41.—Drum record of impact bending test.

allowed to rest upon the specimen and a zero or datum line is drawn. The hammer is then dropped from increasing heights and drum records taken until first failure. The first drop is one inch and the increase is by increments of one inch until a height of ten inches is reached, after which increments of two inches are used until complete failure occurs or 6-inch deflection is secured.

The 50-pound hammer is used when with drops up to 68 inches it is reasonably certain it will produce complete failure or 6-inch deflection in the case of all specimens of a species; for all other species a 100-pound hammer is used.

Results: The tracing on the drum (see Fig. 41) represents the actual deflection of the stick and the subsequent rebounds for each drop. The distance from the lowest point in each case to the datum line is measured and its square in tenths of a square inch entered as an abscissa on cross-section paper, with the height of drop in inches as the ordinate. The elastic limit is that point on the diagram where the square of the deflection begins to increase more rapidly than the height of drop. The difference between the datum line and the final resting point after each drop represents the set the material has received.

The formulæ used in calculating the results of impact tests in bending when the load is applied at the centre up to the elastic limit are as follows:

$$(1) \quad r = \frac{3\,W\,H\,l}{D\,b\,h^2} \qquad (2) \quad E = \frac{F\,S\,l^2}{6\,D\,h} \qquad (3) \quad S = \frac{W\,H}{l\,b\,h}$$

> H = height of drop of hammer, including deflection, inches.
>
> S = modulus of elastic resilience, inch-pounds per cubic inch.
>
> W = weight of hammer, pounds.

Remainder of legend as on page 98.

HARDNESS TEST: ABRASION AND INDENTATION

Abrasion: The machine used by the U. S. Forest Service is a modified form of the Dorry abrasion machine. (See Fig. 42.) Upon the revolving horizontal disk is glued a commercial sand-paper, known as garnet paper, which is commonly employed in factories in finishing wood.

A small block of the wood to be tested is fixed in one clamp and a similar block of some wood chosen as a standard, as sugar maple, at 10 per cent moisture, in the opposite, and held against the same zone of sandpaper by a weight of 26 pounds each. The size of the section under abrasion for each specimen is $2'' \times 2''$.

FIG. 42.—Abrasion machine for testing the wearing qualities of woods.

The conditions for wear are the same for both specimens. The speed of rotation is 68 revolutions a minute.

The test is continued until the standard specimen is worn a specified amount, which varies with the kind of wood under test. A comparison of the wear of the two blocks affords a fair idea of their relative resistance to abrasion.

Another method makes use of a sand blast to abrade the woods and is the one employed in New South Wales.* The apparatus consists essentially of a nozzle through which sand can be propelled at a high velocity against the test specimen by means of a steam jet.

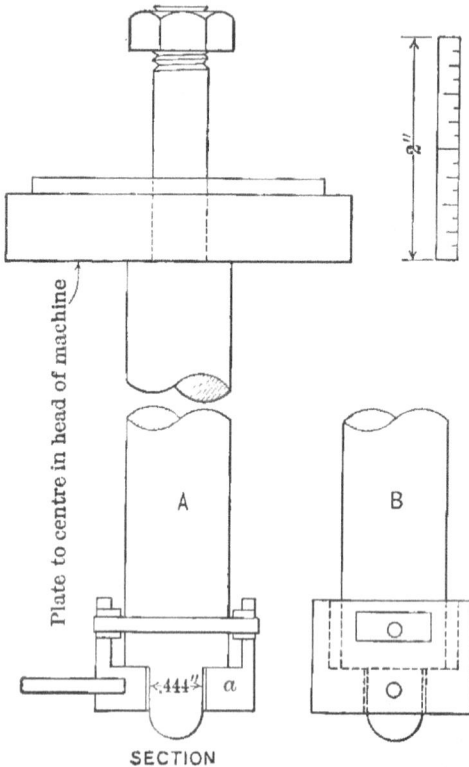

The wood to be tested is cut into blocks 3″ × 3″ × 1′, and these are weighed to the nearest grain just before placing in the apparatus. Steam from the boiler at a pressure of about 43 pounds per square inch is ejected from a nozzle in such a way that particles of fine quartz sand are caught up and thrown violently against the block which is being rotated. Only superheated steam

FIG. 43.—Design of tool for testing the hardness of woods by indentation.

strikes the block, thus leaving the wood dry. The test is continued for two minutes, after which the specimen is removed and immediately weighed.

By comparison with the original weight the loss from

* See Warren, W. H.: The strength, elasticity, and other properties of New South Wales hardwood timbers. Dept. For., N. S. W., Sydney, 1911, pp. 88–95.

abrasion is determined, and by comparison with a certain wood chosen as a standard, a coefficient of wear-resistance can be obtained. The amount of wear will vary more or less according to the surface exposed, and in these tests quarter-sawed material was used with the edge grain to the blast.

Indentation: The tool used for this test consists of a punch with a hemispherical end or steel ball having a diameter of 0.444 inch, giving a surface area of one-fourth square inch. It is fitted with a guard plate, which works loosely until the penetration has progressed to a depth of 0.222 inch, whereupon it tightens. (See Fig. 43.) The effect is that of sinking a ball half its diameter into the specimen. This apparatus is fitted into the movable head of the static testing machine.

The wood to be tested is cut square with the grain into rectangular blocks measuring $2'' \times 2'' \times 6''$. A block is placed on the platform and the end of the punch forced into the wood at the rate of 0.25 inch per minute. The operator keeps moving the small handle of

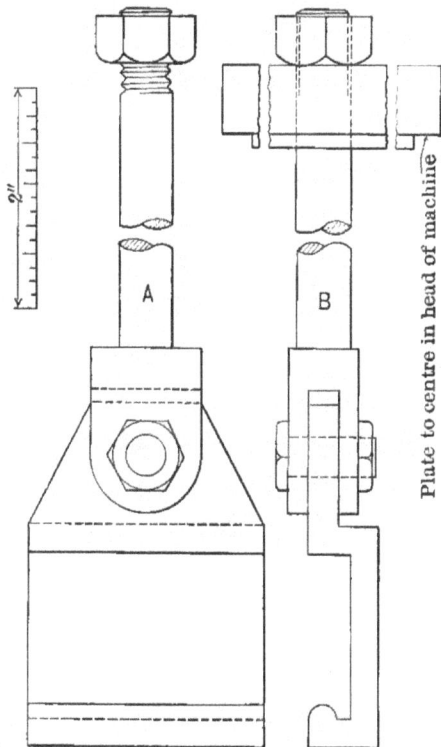

FIG. 44.—Design of tool for cleavage test.

the guard plate back and forth until it tightens. At this instant the load is read and recorded.

Two penetrations each are made on the tangential and radial surfaces, and one on each end of every specimen tested.

In choosing the places on the block for the indentations, effort should be made to get a fair average of heartwood and sapwood, fine and coarse grain, early and late wood.

Another method of testing by indentation involves the use

of a right-angled cone instead of a ball. For details of this test as used in New South Wales see *loc. cit.*, pp. 86–87.

CLEAVAGE TEST

A static testing machine and a special cleavage testing device are required. (See Fig. 44.) The latter consists essentially of two hooks, one of which is suspended from the centre of the top of the cage, the other extended above the movable head.

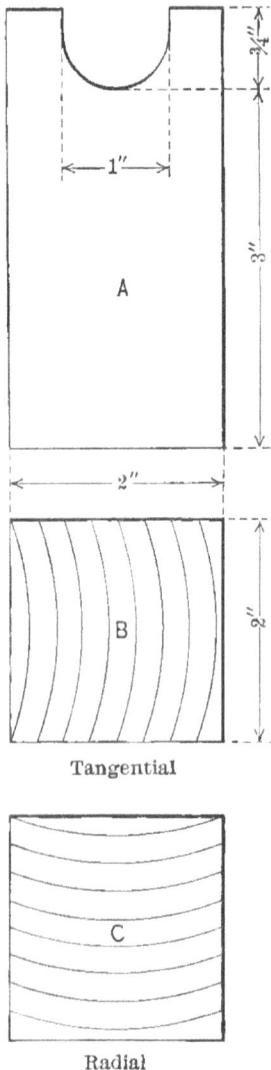

The specimens are $2'' \times 2'' \times 3.75''$. At one end a one-inch hole is bored, with its centre equidistant from the two sides and 0.25 inch from the end. (See Fig. 45.) This makes the cross section to be tested $2'' \times 3''$. Some of the blocks are cut radially and some tangentially, as indicated in the figure.

The free ends of the hooks are fitted into the notch in the end of the specimen. The movable head of the machine is then made to descend at the rate of 0.25 inch per minute, pulling apart the hooks and splitting the block. The maximum load only is taken and the result expressed in pounds per square inch of width. A piece one-half inch thick is split off parallel to the failure and used for moisture determination.

FIG. 45.—Design of cleavage test specimen.

TENSION TEST PARALLEL TO THE GRAIN

Since the tensile strength of wood parallel to the grain is greater than the compressive strength, and exceedingly greater than the shearing strength, it is very difficult to make satisfactory tension tests, as the head and shoulders of the test specimen (which is subjected to both compression and shear) must be stronger than the portion subjected to a pure tensile stress.

Various designs of test specimens have been made. The one first employed by the Division of Forestry * was prepared as follows: Sticks were cut measuring $1.5'' \times 2.5'' \times 16''$. The thickness at the centre was then reduced to three-eighths of an

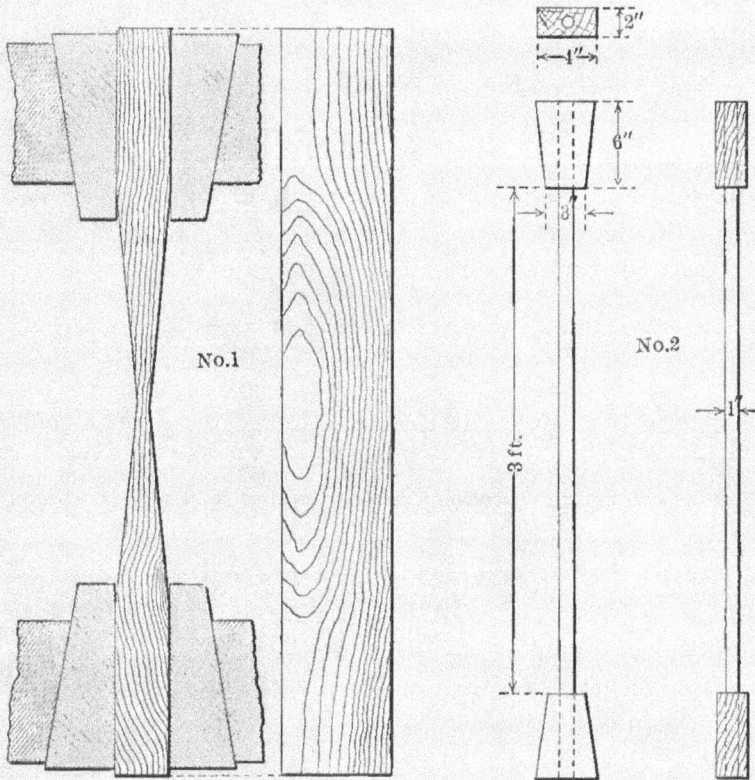

FIG. 46.—Designs of tension test specimens used in United States.

inch by cutting out circular segments with a band saw. This left a breaking section of $2.5'' \times 0.375''$. Care was taken to cut the specimen as nearly parallel to the grain as possible, so that its failure would occur in a condition of pure tension. The specimen was then placed between the plane wedge-shaped steel grips of the cage and the movable head of the static machine and pulled in two. Only the maximum load was recorded. (See Fig. 46, No. 1.)

* Bul. No. 8: Timber physics. Part II., 1893, p. 7.

The difficulty of making such tests compared with the minor importance of the results is so great that they are at present omitted by the U. S. Forest Service. A form of specimen is suggested, however, and is as follows: " A rod of wood about one inch in diameter is bored by a hollow drill from the stick to be tested. The ends of this rod are inserted and glued in corresponding holes in permanent hardwood wedges. The specimen is then submitted to the ordinary tension test. The broken ends are punched from the wedges." * (See Fig. 46, No. 2.)

The form used by the Department of Forestry of New South Wales † is as shown in Fig. 47. The specimen has a total length of 41 inches and is circular in cross section. On each end is a head 4 inches in diameter and 7 inches long. Below each head is a

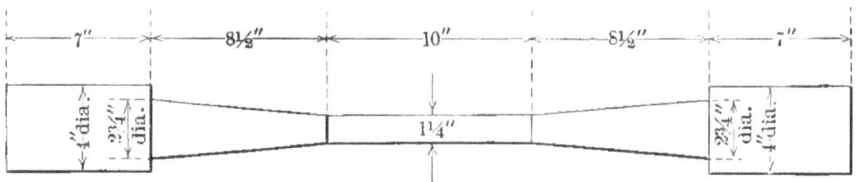

FIG. 47.—Design of tension test specimen used in New South Wales.

shoulder 8.5 inches long, which tapers from a diameter of 2.75 inches to 1.25 inches. In the middle is a cylindrical portion 1.25 inches in diameter and 10 inches long.

In making the test the specimen is fitted in the machine, and an extensometer attached to the middle portion and arranged to record the extension between the gauge points 8 inches apart. The area of the cross section then is 1.226 square inches, and the tensile strength is equal to the total breaking load applied divided by this area.

TENSION TEST AT RIGHT ANGLES TO THE GRAIN

A static testing machine and a special testing device (see Fig. 48) are required. The latter consists essentially of two double

---

* Cir. 38: Instructions to engineers of timber tests, 1906, p. 24.
† Warren, W. H.: The strength, elasticity, and other properties of New South Wales hardwood timbers, 1911, pp. 58–62.

hooks or clamps, one of which is suspended from the centre of the top of the cage, the other extended above the movable head.

The specimens are $2'' \times 2'' \times 2.5''$. At each end a one-inch

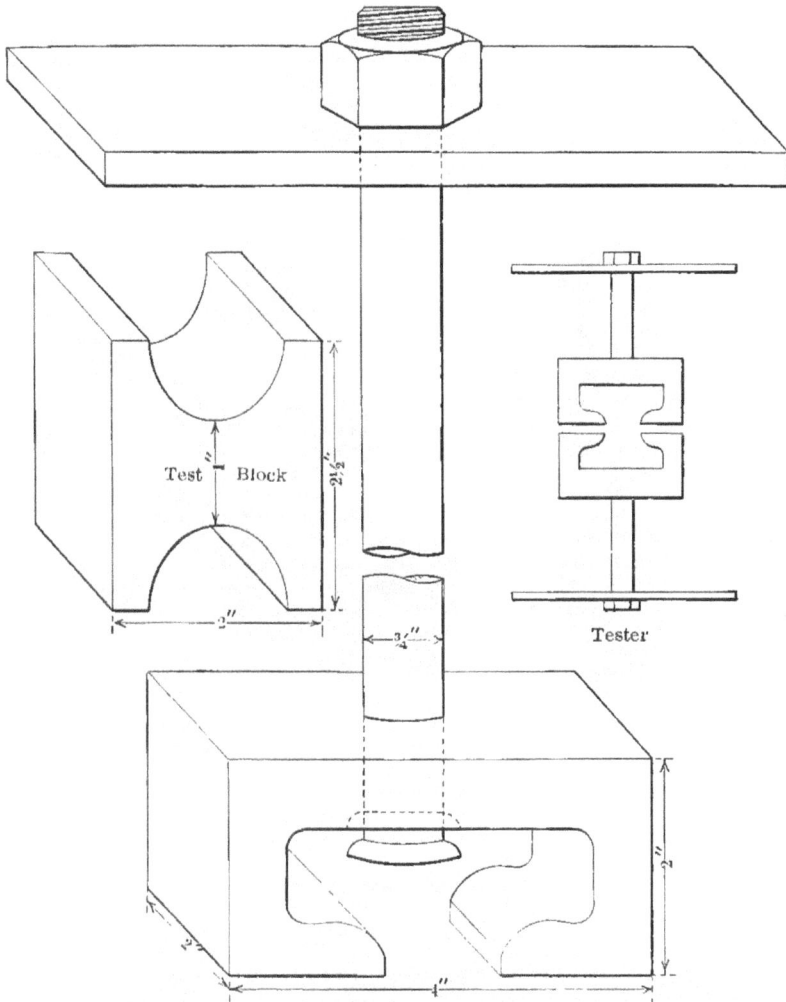

FIG. 48.—Design of tool and specimen for testing tension at right angles to the grain.

hole is bored with its centre equidistant from the two sides and 0.25 inch from the ends. This makes the cross section to be tested $1'' \times 2''$.

The free ends of the clamps are fitted into the notches in the ends of the specimen. The movable head of the machine is then made to descend at the rate of 0.25 inch per minute, pulling the specimen in two at right angles to the grain. The maximum load only is taken and the result expressed in pounds per

Photo by U. S. Forest Service.

FIG. 49.—Making a torsion test on hickory.

inch of width. A piece one-half inch thick is split off parallel to the failure and used for moisture determination.

TORSION TEST *

Apparatus: The torsion test is made in a Riehle-Miller torsional testing machine or its equivalent. (See Fig. 49.)

Preparation of material: The test pieces are cylindrical, 1.5 inches in diameter and 18 inches gauge length, with squared ends 4 inches long joined to the cylindrical portion with a fillet. The dimensions are carefully measured, and the usual data obtained in regard to the rate of growth, proportion of late wood, location

---

Wood is so seldom subjected to a pure stress of this kind that the torsion test is usually omitted.

and kind of defects. The weight of the cylindrical portion of the specimen is obtained after the test.

Making the test: After the specimen is fitted in the machine the load is applied continuously at the rate of 22° per minute. A troptometer is used in measuring the deformation. Readings are made until failure occurs, the points being entered on the cross-section paper. The character of the failure is described. Moisture determinations are made by the disk method.

Results: The conditions of ultimate rupture due to torsion appear not to be governed by definite mathematical laws; but where the material is not overstrained, laws may be assumed which are sufficiently exact for practical cases. The formulæ commonly used for computations are as follows:

$$(1) \quad T = \frac{5.1\,M}{c^3} \qquad\qquad (2)\; G = \frac{114.6\,T\,f}{a\,c}$$

a = angle measured by troptometer at elastic limit, in degrees.

c = diameter of specimen, inches.

f = gauge length of specimen, inches.

G = modulus of elasticity in shear across the grain, pounds per square inch.

M = moment of torsion at elastic limit, inch-pounds.

T = outer fibre torsional stress at elastic limit, pounds per square inch.

SPECIAL TESTS

Spike-pulling Test

Spike-pulling tests apply to problems of railroad maintenance, and the results are used to compare the spike-holding powers of various woods, both untreated and treated with different preservatives, and the efficiency of various forms of spikes. Special tests are also made in which the spike is subjected to a transverse load applied repetitively by a blow.

For details of tests and results see:

Cir. 38,   U.S.F.S.: Instructions to engineers of timber tests, p. 26.

Cir. 46,   U.S.F.S.: Holding force of railroad spikes in wooden ties.

Bul. 118, U.S.F.S.: Prolonging the life of cross-ties, pp. 37–40.

Packing Boxes

Special tests on the strength of packing boxes of various woods have been made by the U. S. Forest Service to determine the merits of different kinds of woods as box material with the view of substituting new kinds for the more expensive ones now in use. The methods of tests consisted in applying a load along the diagonal of a box, an action similar to that which occurs when a box is dropped on one of its corners. The load was measured at each one-fourth inch in deflection, and notes were made of the primary and subsequent failures.

For details of tests and results, see:

Cir. 47, U.S.F.S.: Strength of packing boxes of various woods.

Cir. 214, U.S.F.S.: Tests of packing boxes of various forms.

Vehicle and Implement Woods

Tests were made by the U. S. Forest Service to obtain a better knowledge of the mechanical properties of the woods at present used in the manufacture of vehicles and implements and of those which might be substituted for them. Tests were made upon the following materials: hickory buggy spokes (see Fig. 5, page 11); hickory and red oak buggy shafts; wagon tongues; Douglas fir and southern pine cultivator poles.

Details of the tests and results may be found in:

Cir. 142, U.S.F.S.: Tests on vehicle and implement woods.

Cross-arms

In tests by the U. S. Forest Service on cross-arms a special apparatus was devised in which the load was distributed along the arm as in actual practice. The load was applied by rods passing through the pinholes in the arms. Nuts on these rods pulled down on the wooden bearing-blocks shaped to fit the upper side of the arm. The lower ends of these rods were attached to a system of equalizing levers, so arranged that the load at each pinhole would be the same. In all the tests the load was applied vertically by means of the static machine.

See Cir. 204, U.S.F.S.: Strength tests of cross-arms.

Other Tests

Many other kinds of tests are made as occasion demands. One kind consists of barrels and liquid containers, match-boxes, and explosive containers. These articles are subjected to shocks such as they would receive in transit and in handling, and also to hydraulic pressure.

One of the most important tests from a practical standpoint is that of built-up structures such as compounded beams composed of small pieces bolted together, mortised joints, wooden trusses, etc. Tests of this kind can best be worked out according to the specific requirements in each case.

APPENDIX

SAMPLE WORKING PLAN OF THE
U. S. FOREST SERVICE

MECHANICAL PROPERTIES OF WOODS GROWN IN THE UNITED STATES

Working Plan No. 124

PURPOSE OF WORK

IT is the general purpose of the work here outlined to provide:

(a) Reliable data for comparing the mechanical properties of various species;

(b) Data for the establishment of correct strength functions or working stresses;

(c) Data upon which may be based analyses of the influence on the mechanical properties of such factors as:

Locality;

Distance of timber from the pith of the tree;

Height of timber in the tree;

Change from the green to the air-dried condition, etc.

The mechanical properties which will be considered and the principal tests used to determine them are as follows:

Strength and stiffness—

Static bending;

Compression parallel to grain;

Compression perpendicular to grain;

Shear.

Toughness—

Impact bending;

Static bending;

Work to maximum load and total work.

127

Cleavability—
  Cleavage test.
Hardness—
  Modification of Janka ball test for surface hardness.

Selection and Number of Trees

The material will be from trees selected in the forest by one qualified to determine the species. From each locality, three to five dominant trees of merchantable size and approximately average age will be so chosen as to be representative of the dominant trees of the species. Each species will eventually be represented by trees from five to ten localities. These localities will be so chosen as to be representative of the commercial range of the species. Trees from one to three localities will be used to represent each species until most of the important species have been tested.

The 16-foot butt log will be taken from each tree selected and the entire merchantable bole of one average tree for each species.

Field Notes and Shipping Instructions

Field notes as outlined in Form —a Shipment Description, Manual of the Branch of Products, will be fully and carefully made by the collector. The age of each tree selected will be recorded and any other information likely to be of interest or importance will also be made a part of these field notes. Each log will have the bark left on. It will be plainly marked in accordance with directions given under Detailed Instructions. All material will be shipped to the laboratory immediately after being cut. No trees will be cut until the collector is notified that the laboratory is ready to receive the material.

Part of Tree to be Tested

(a) For determining the value of tree and locality and the influence on the mechanical properties of distance from the pith, a 4-foot bolt will be cut from the top end of each 16-foot butt log.

(b) For investigating the variation of properties with the height of timber in the tree, all the logs from one average tree will be used.

(c) For investigating the effect of drying the wood, the bolt next below that provided for in (a) will be used in the case of one tree from each locality.

Marking and Grouping of Material

The marking will be standard except as noted. Each log will be considered a " piece." The piece numbers will be plainly marked upon the butt end of each log by the collector. The north side of each log will also be marked.

When only one bolt from a tree is used it will be designated by the number of the log from which it is cut. Whenever more than one bolt is taken from a tree, each 4-foot bolt or length of trunk will be given a letter (mark), a, b, c, etc., beginning at the stump.

All bolts will be sawed into $2\frac{1}{2}'' \times 2\frac{1}{2}''$ sticks and the sticks marked according to the sketch, Fig. 50. The letters N, E, S, and W indicate the cardinal points when known; when these are unknown, H, K, L, and M will be used. Thus, N5, K8, S7, M4 are stick numbers, the letter being a part of the stick number.

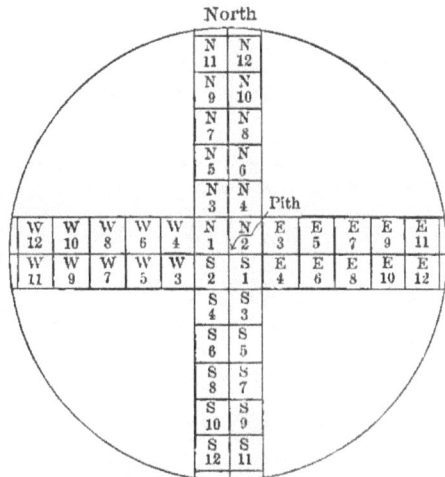

Fig. 50.—Method of cutting and marking test specimens.

Only straight-grained specimens, free from defects which will affect their strength, will be tested.

Care of Material

No material will be kept in the bolt or log long enough to be damaged or disfigured by checks, rot, or stains.

Green material: The material to be tested green will be kept in a green state by being submerged in water until near the time of test. It will then be surfaced, sawed to length, and stored in damp sawdust at a temperature of 70° F. (as nearly as practicable) until time of test. Care should be taken to avoid as much as possible the storage of green material in any form.

Air-dry material: The material to be air-dried will be cut into sticks $2\frac{1}{2}'' \times 2\frac{1}{2}'' \times 4'$. The ends of these sticks will be paraffined to prevent checking. This material will be so piled as to leave an air space of at least one-half inch on each side of each stick, and in such a place that it will be protected from sunshine, rain, snow, and moisture from the ground. The sticks will be surfaced and cut to length just previous to test.

Order of Tests

The order of tests in all cases will be such as to eliminate so far as possible from the comparisons the effect of changes of condition of the specimens due to such factors as storage and weather conditions.

The material used for determining the effect of height in tree will be tested in such order that the average time elapsing from time of cutting to time of test will be approximately the same for all bolts from any one tree.

Tests on Green Material

The tests on all bolts, except those from which a comparison of green and dry timber is to be gotten, will be as follows:

Static bending: One stick from each pair. A pair consists of two adjacent sticks equidistant from the pith, as $N7$ and $N8$, or $H5$ and $H6$.

Impact bending: Four sticks; one to be taken from near the pith; one from near the periphery; and two representative of the cross section.

Compression parallel to grain: One specimen from each stick. These will be marked " 1 " in addition to the number of the stick from which they are taken.

Compression perpendicular to grain: One specimen from each of 50 per cent of the static bending sticks. These will be marked

" 2 " in addition to the number of the stick from which they are cut.

Hardness: One specimen from each of the other 50 per cent of the static bending sticks. These specimens will be marked " 4."

Shear: Six specimens from sticks not tested in bending or from the ends cut off in preparing the bending specimens. Two specimens will be taken from near the pith; two from near the periphery; and two that are representative of the average growth. One of each two will be tested in radial shear and the other in tangential shear. These specimens will have the mark " 3."

Cleavage: Six specimens chosen and divided just as those for shearing. These specimens will have the mark " 5." (For sketches showing radial and tangential cleavage, see Fig. 45, page 118.)

When it is impossible to secure clear specimens for all of the above tests, tests will have precedence in the order in which they are named.

Tests to Determine the Effect of Air-drying

These tests will be made on material from the adjacent bolts mentioned in " c " under Part of Tree to be Tested. Both bolts will be cut as outlined above. One-half the sticks from each bolt will be tested green, the other half will be air-dried and tested. The division of green and air-dry will be according to the following scheme:

STICK NUMBERS

Lower bolt,	1,	4, 5,	8, 9,		Tested
				etc.	green
Upper bolt,	2, 3,	6, 7,	10,		

Lower bolt,	2, 3,	6, 7,	10,		Air-dried
				etc.	and
Upper bolt,	1,	4, 5,	8, 9,		tested

All green sticks from these two bolts will be tested as if they were from the same bolt and according to the plan previously outlined for green material from single bolts. The tests on the air-dried material will be the same as on the green except for the difference of seasoning.

The material will be tested at as near 12 per cent moisture as is practicable. The approximate weight of the air-dried specimens at 12 per cent moisture will be determined by measuring while green 20 per cent of the sticks to be air-dried and assuming their dry gravity to be the same as that of the specimens tested green. This 20 per cent will be weighed as often as is necessary to determine the proper time of test.

Methods of Test

All tests will be made according to Circular 38 except in case of conflict with the instructions given below:

Static bending: The tests will be on specimens 2″ × 2″ × 30″ on 28-inch span. Load will be applied at the centre.

In all tests the load-deflection curve will be carried to or beyond the maximum load. In one-third of the tests the load-deflection curve will be continued to 6-inch deflection, or till the specimen fails to support a 200-pound load. Deflection readings for equal increments of load will be taken until well past the elastic limit, after which the scale beam will be kept balanced and the load read for each 0.1-inch deflection. The load and deflection at first failure, maximum load and points of sudden change, will be shown on the curve sheet even if they do not occur at one of the regular load or deflection increments.

Impact bending: The impact bending tests will be on specimens of the same size as those used in static bending. The span will be 28 inches.

The tests will be by increment drop. The first drop will be 1 inch and the increase will be by increments of 1 inch till a height of 10 inches is reached, after which increments of 2 inches will be used until complete failure occurs or 6-inch deflection is secured.

A 50-pound hammer will be used when with drops up to 68 inches it is practically certain that it will produce complete failure or 6-inch deflection in the case of all specimens of a species. For all other species, a 100-pound hammer will be used.

In all cases drum records will be made until first failure. Also the height of drop causing complete failure or 6-inch deflection will be noted.

Compression parallel to grain: This test will be on specimens

$2'' \times 2'' \times 8''$ in size.   On 20 per cent of these tests load-compression curves for a 6-inch centrally located gauge length will be taken. Readings will be continued until the elastic limit is well passed. The other 80 per cent of the tests will be made for the purpose of obtaining the maximum load only.

Compression perpendicular to grain: This test will be on specimens $2'' \times 2'' \times 6''$ in size.   The bearing plates will be 2 inches wide.   The rate of descent of the moving head will be 0.024 inch per minute.   The load-compression curve will be plotted to 0.1 inch compression and the test will then be discontinued.

Hardness: The tool shown in Fig. 43, page 116 (an adaptation of the apparatus used by the German investigator, Janka) will be used.   The rate of descent of the moving head will be 0.25 inch per minute.   When the penetration has progressed to the point at which the plate " *a* " becomes tight, due to being pressed against the wood, the load will be read and recorded.

Two penetrations will be made on a tangential surface, two on a radial, and one on each end of each specimen tested.   The choice between the two radial and between the two tangential surfaces and the distribution of the penetrations over the surfaces will be so made as to get a fair average of heart and sap, slow and fast growth, and spring and summer wood.   Specimens will be $2'' \times 2'' \times 6''$.

Shear: The tests will be made with a tool slightly modified from that shown in Circular 38.   The speed of descent of head will be 0.015 inch per minute.   The only measurements to be made are those of the shearing area.   The offset will be $\frac{1}{8}$ inch.   Specimens will be $2'' \times 2'' \times 2\frac{1}{2}''$ in size.   (For definition of offset and form of test specimen, see Fig. 38, page 108.)

Cleavage: The cleavage tests will be made on specimens of the form and size shown in Fig. 45, page 118.   The apparatus will be as shown in Fig. 44.   The maximum load only will be taken and the result expressed in pounds per inch of width.   The speed of the moving head will be 0.25 inch per minute.

Moisture Determinations

Moisture determinations will be made on all specimens tested except those to be photographed or kept for exhibit.   A 1-inch disk will be cut from near the point of failure of bending and compression parallel specimens, from the portion under the plate in the case of the compression perpendicular specimens, and from the centre of the hardness test specimens. The beads from the shear specimens will be used as moisture disks. In the case of the cleavage specimens a piece ½ inch thick will be split off parallel to the failure and used as a moisture disk.

RECORDS

All records will be standard.

PHOTOGRAPHS

Cross Sections

Just before cutting into sticks, the freshly cut end of at least one bolt from each tree will be photographed. A scale of inches will be shown in this photograph.

Specimens

Three photographs will be made of a group consisting of four $2'' \times 2'' \times 30''$ specimens chosen from the material from each locality. Two of these specimens will be representative of average growth, one of fast and one of slow growth. These photographs will show radial, tangential, and end surfaces for each specimen.

Failures

Typical and abnormal failures of material from each site will be photographed.

Disposition of Material

The specimens photographed to show typical and abnormal failures will be saved for purposes of exhibit until deemed by the person in charge of the laboratory to be of no further value.

SHRINKAGE AND SPECIFIC GRAVITY
Appendix to Working Plan 124

It is the purpose of this work to secure data on the shrinkage and specific gravity of woods tested under Project 124. The figures to be obtained are for use as average working values rather than as the basis for a detailed study of the principles involved.

MATERIAL

The material will be taken from that provided for mechanical tests.

RADIAL AND TANGENTIAL SHRINKAGE

Specimens

Preparation: Two specimens 1 inch thick, 4 inches wide, and 1 inch long will be obtained from near the periphery of each " *d* " bolt. These will be cut from the sector-shaped sections left after securing the material for the mechanical tests or from disks cut from near the end of the bolt. They will be taken from adjoining pieces chosen so that the results will be comparable for use in determining radial and tangential shrinkage. (When a disk is used, care must be taken that it is green and has not been affected by the shrinkage and checking near the end of the bolt.)

One of these specimens will be cut with its width in the radial direction and will be used for the determination of radial shrinkage. The other will have its width in the tangential direction and will be used for tangential shrinkage. These specimens will not be surfaced.

Marking: The shrinkage specimens will retain the shipment and piece numbers and marks of the bolts from which they are taken, and will have the additional mark 7*R* or 7*T* according as their widths are in the radial or tangential direction.

Shrinkage measurements: The shrinkage specimens will be carefully weighed and measured soon after cutting. Rings per inch, per cent sap, and per cent summer wood will be measured. They will then be air-dried in the laboratory to constant weight, and afterward oven-dried at 100 ° C. (212° F.), when they will again be weighed and measured.

VOLUMETRIC SHRINKAGE AND SPECIFIC GRAVITY

Specimens

Selection and preparation: Four 2″ × 2″ × 6″ specimens will be cut from the mechanical test sticks of each " *d* " bolt; also from each of the composite bolts used in getting a comparison of green and air-dry. One of these specimens will be taken from near the pith and one from near the periphery; the other two will be representative of the average growth of the bolt. The sides of these specimens will be surfaced and the ends smooth sawn.

Marking: Each specimen will retain the shipment, piece, and stick numbers and mark of the stick from which it is cut, and will have the additional mark " *S.*"

Manipulation: Soon after cutting, each specimen will be weighed and its volume will be determined by the method described below. The rings per inch and per cent summer wood, where possible, will be determined, and a carbon impression of the end of the specimen made. It will then be air-dried in the laboratory to a constant weight and afterward oven-dried at 100° C. When dry, the specimen will be taken from the oven, weighed, and a carbon impression of its end made. While still warm the specimen will be dipped in hot paraffine. The volume will then be determined by the following method:

On one pan of a pair of balances is placed a container having in it water enough for the complete submersion of the test specimen. This container and water is balanced by weights placed on the other scale pan. The specimen is then held completely submerged and not touching the container while the scales are again balanced. The weight required to balance is the weight of water displaced by the specimen, and hence if in grams is numerically equal to the volume of the specimen in cubic centimetres.

A diagrammatic sketch of the arrangement of this apparatus is shown in Fig. 51.

Air-dry specimens will be dipped in water and then wiped dry after the first weighing and just before being immersed for weighing

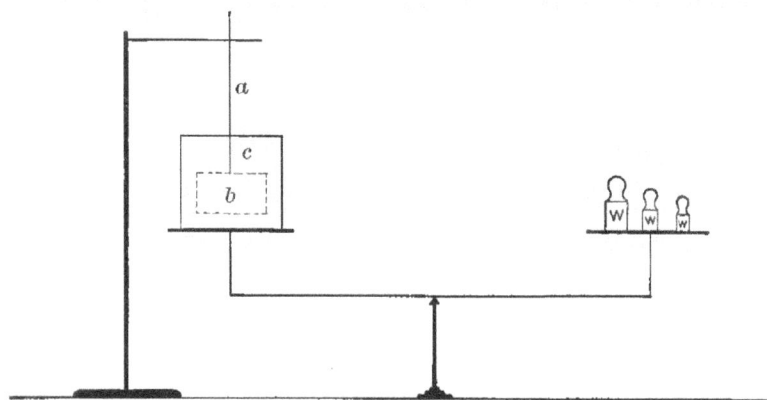

Fig. 51.—Diagram of specific gravity apparatus, showing a balance with container (c) filled with water in which the test block (b) is held submerged by a light rod (a) which is adjustable vertically and provided with a sharp point to be driven into the specimen.

their displacement. All displacement determinations will be made as quickly as possible in order to minimize the absorption of water by the specimen.

STRENGTH VALUES FOR STRUCTURAL TIMBERS

(From Cir. 189, U. S. Forest Service)

The following tables bring together in condensed form the average strength values resulting from a large number of tests made by the Forest Service on the principal structural timbers of the United States. These results are more completely discussed in other publications of the Service, a list of which is given on pages 157–159.

The tests were made at the laboratories of the U. S. Forest Service, in coöperation with the following institutions: Yale Forest School, Purdue University, University of California, University of Oregon, University of Washington, University of Colorado, and University of Wisconsin.

Tables XVIII and XIX give the average results obtained from tests on green material, while Tables XX and XXI give average results from tests on air-seasoned material. The small specimens, which were invariably 2″ × 2″ in cross section, were free from defects such as knots, checks, and cross grain; all other specimens were representative of material secured in the open market. The relation of stresses developed in different structural forms to those developed in the small clear specimens is shown for each factor in the column headed " Ratio to 2″ × 2″." Tests to determine the mechanical properties of different species are often confined to small, clear specimens. The ratios included in the tables may be applied to such results in order to approximate the strength of the species in structural sizes, and containing the defects usually encountered, when tests on such forms are not available.

A comparison of the results of tests on seasoned material with those from tests on green material shows that, without exception, the strength of the 2″ × 2″ specimens is increased by lowering the moisture content, but that increase in strength of other sizes is much more erratic. Some specimens, in fact, show an apparent loss in strength due to seasoning. If structural timbers are seasoned slowly, in order to avoid excessive checking, there should be an increase in their strength. In the light of these facts it is not safe to base working stresses on results secured from any but green material. For a discussion of factors of safety and safe working stresses for structural timbers see the Manual of the American Railway Engineering Association, Chicago, 1911. A table from that publication, giving working unit stresses for structural timber, is reproduced on page 144 of this book.

TABLE XVIII

BENDING TESTS ON GREEN MATERIAL

Species	Sizes		Number of tests	Per cent of moisture	Rings per inch	F. S. at E.L.		M. of R.		M. of E.		Calculated shear	
	Cross section	Span				Average per square inch	Ratio to 2" by 2"	Average per square inch	Ratio to 2" by 2"	Average per square inch	Ratio to 2" by 2"	Average per square inch	Ratio to 2" by 2"
	Inches	Ins.				Lbs.		Lbs.		1 000 lbs.		Lbs.	
Longleaf pine.....	12 by 12	138	4	28.6	9.7	4,029	0.83	6,710	0.74	1,523	0.99	261	0 86
	10 by 16	168	4	26.8	16.7	4,193	.85	6,453	.71	1,626	1.05	306	1 01
	8 by 16	156	7	28.4	14.6	3,147	.64	5,439	.60	1,368	89	390	1 29
	6 by 16	132	1	40.3	21.8	4,120	.83	6,460	.71	1,190	.77	378	1 25
	6 by 10	180	1	31.0	6.2	3,580	.72	6,500	.72	1,412	.92	175	58
	6 by 8	180	2	27.0	8.2	3,735	.75	5,745	.63	1,282	.83	121	.40
	2 by 2	30	15	33.9	14.1	4,950	1.00	9,070	1.00	1,540	1.00	303	1.00
Douglas fir.......	8 by 16	180	191	31.5	11.0	3,968	.76	5,983	.72	1,517	.95	269	.81
	5 by 8	180	84	30.1	10.8	3,693	.71	5,178	.63	1,533	96	172	.52
	2 by 12	180	27	35.7	20.3	3,721	.71	5,276	.64	1,642	1.03	256	.77
	2 by 10	180	26	32.9	21.6	3,160	.60	4,699	.57	1,593	1 00	189	.57
	2 by 8	180	29	33.6	17.6	3,593	.69	5,352	.65	1,607	1 01	171	.51
	2 by 2	24	568	30.4	11.6	5,227	1.00	8,280	1 00	1,597	1.00	333	1 00
Douglas fir (fire-killed)........	8 by 16	180	30	36.8	10 9	3,503	.80	4,994	.64	1,531	.94	330	1.19
	2 by 12	180	32	34.2	17.7	3,489	.80	5,085	.66	1,624	99	247	.89
	2 by 10	180	32	38.9	18.1	3,851	.88	5,359	.69	1,716	1.05	216	.78
	2 by 8	180	31	37.0	15 7	3,403	.78	5,305	.68	1,676	1 02	169	61
	2 by 2	30	290	33.2	17.2	4,360	1.00	7,752	1.00	1,636	1 00	277	1.00
Shortleaf pine.....	8 by 16	180	12	39.5	12.1	3,185	.73	5,407	.70	1,438	1 03	362	1.40
	8 by 14	180	12	45.8	12 7	3,234	.74	5,781	.75	1,494	1 07	338	1.31
	8 by 12	180	24	52.2	11.8	3,265	.75	5,503	.71	1,480	1.06	277	1.07
	5 by 8	180	24	47.8	11.5	3,519	.81	5,732	.74	1,485	1.06	185	.72
	2 by 2	30	254	51.7	13.6	4,350	1 00	7,710	1.00	1,395	1.00	258	1.00
Western larch.....	8 by 16	180	32	51.0	25.3	3,276	.77	4,632	.64	1,272	.97	298	1.11
	8 by 12	180	30	50.3	23.2	3,376	.79	5,286	.73	1,331	1.02	254	.94
	5 by 8	180	14	56.0	25.6	3,528	.83	5,331	.74	1,432	1.09	169	.63
	2 by 2	28	189	46.2	26.2	4,274	1.00	7,251	1.00	1,310	1.00	269	1.00
Loblolly pine.....	8 by 16	180	17	55.8	6.1	3,094	.75	5,394	.69	1,406	.98	383	1.44
	5 by 12	180	94	60.9	5.9	3,030	.74	5,028	64	1,383	96	221	.83
	2 by 2	28	44	70.9	5.4	4,100	1.00	7,870	1.00	1,440	1 00	265	1.00
Tamarack........	6 by 12	162	15	57.6	16.6	2,914	.75	4,500	.66	1,202	1 05	255	1.11
	4 by 10	162	15	43.5	11.4	2,712	.70	4,611	.68	1,238	1.08	209	.91
	2 by 2	30	82	38.8	14.0	3,875	1.00	6,820	1.00	1,141	1 00	229	1.00
Western hemlock..	8 by 16	180	39	42.5	15.6	3,516	.80	5,296	73	1,445	1 01	261	.92
	2 by 2	28	52	51.8	12.1	4,406	1.00	7,294	1.00	1,428	1 00	284	1.00
Redwood........	8 by 16	180	14	86.5	19.9	3,734	.79	4,492	.64	1,016	96	300	1.21
	6 by 12	180	14	87.3	17.8	3,787	.80	4,451	.64	1,068	1 00	224	.90
	7 by 9	180	14	79.8	16.7	4,412	.93	5,279	.76	1,324	1 25	199	.80
	3 by 16	180	13	86.1	23.7	3,506	.74	4,364	.62	947	.89	255	1.03
	2 by 12	180	12	70.9	18.6	3,100	.65	3,753	.54	1,052	.99	187	.75
	2 by 10	180	13	55.8	20.0	3,285	.69	4,079	.58	1,107	1.04	169	68
	2 by 8	180	13	63.8	21.5	2,989	.63	4,063	.58	1,141	1.08	134	.54
	2 by 2	28	157	75.5	19 1	4,750	1.00	6,980	1.00	1,061	1.00	248	1.00
Norway pine......	6 by 12	162	15	50.3	12.5	2,305	.82	3,572	.69	987	1.03	201	1.17
	4 by 12	162	18	47.9	14.7	2,648	.94	4,107	.79	1,255	1.31	238	1.38
	4 by 10	162	16	45.7	13.3	2,674	95	4,205	.81	1,306	1.36	198	1.15
	2 by 2	30	133	32.3	11.4	2,808	1.00	5,173	1.00	960	1.00	172	1.00
Red spruce.......	2 by 10	144	14	32 5	21.9	2,394	.66	3,566	.60	1,180	1.02	181	.80
	2 by 2	26	60	37.3	21.3	3,627	1.00	5,900	1.00	1,157	1.00	227	1.00
White spruce.....	2 by 10	144	16	40.7	9.3	2,239	.72	3,288	.63	1,081	1.08	166	.83
	2 by 2	26	83	58.3	10.2	3,090	1 00	5,185	1.00	998	1.00	199	1.00

TABLE XIX

COMPRESSION AND SHEAR TESTS ON GREEN MATERIAL

Species	Compression ‖ to grain						Compression ⊥ to grain					Shear		
	Size of specimen	Number of tests	Per cent of moisture	Cr. str. at E. L., per square inch	M. of E., per square inch	Cr. str. at max. ld., per square inch	Stress area	Height	Number of tests	Per cent of moisture	Cr. str. at E. L., per square inch	Number of tests	Per cent of moisture	Shear strength
	Inches			Lbs.	1,000 lbs.	Lbs.	Inches	In.			Lbs.			Lbs.
Longleaf pine....	4 by 4	46	26.3	3,480	....	4,800	4 by 4	4	22	25.3	568	44	21.8	973
	2 by 2	14	34.7	....	....	4,400								
Douglas fir......	6 by 6	515	30.7	2,780	1,181	3,500	4 by 8	16	259	30.3	570	531	29.7	765
	5 by 6	170	30.9	2,720	2,123	3,490	....	..	...	....	...	...	....	...
	2 by 2	902	29.8	3,500	1,925	4,030	....	..	...	....	...	...	....	...
Douglas fir (fire-killed)....	6 by 6	108	34.8	2,620	1,801	3,290	6 by 8	16	24	33.7	368	77	35.8	631
	2 by 2	204	37.9	....	....	3,430								
Shortleaf pine...	6 by 6	95	41.2	2,514	1,565	3,436	5 by 8	16	12	37.7	361	179	47.0	704
	5 by 8	23	43.5	2,241	1,529	3,423	5 by 8	14	12	42.8	366	...	....	...
	2 by 2	281	51.4	....	....	3,570	5 by 8	12	24	53.0	325	...	....	...
	....	...	....	....	....	....	5 by 5	8	24	47.0	344	...	....	...
	....	...	....	....	....	....	2 by 2	2	277	48.5	400			
Western larch...	6 by 6	107	49.1	2,675	1,575	3,510	6 by 8	16	22	43.6	417	179	40.7	700
	2 by 2	491	50.6	3,026	1,545	3,696	6 by 8	12	20	40.2	416	...	....	...
	....						4 by 6	6	53	52.8	478	...	....	...
							4 by 4	4	30	50.4	472	...	....	...
Loblolly pine....	8 by 8	14	63.4	1,560	365	2,140	8 by 4	8	16	67.2	392	121	83.2	630
	4 by 8	18	60.0	2,430	691	3,560	4 by 4	8	38	44.6	546	...	....	...
	2 by 2	53	74.0	....	....	3,240	....	..	...	....	...			
Tamarack......	6 by 7	4	49.9	2,332	1,432	3,032	....	..	...	....	...	24	39.2	668
	4 by 7	6	27.7	2,444	1,334	3,360	....	..	...	....	...	...	....	...
	2 by 2	165	36.8	....	....	3,190	....	..	....	....	...			
Western hemlock........	6 by 6	82	46.6	2,905	1,617	3,355	6 by 4	6	30	48.7	434	54	65.7	630
	2 by 2	131	55.6	2,938	1,737	3,392								
Redwood......	6 by 6	34	83.6	3,194	1,240	3,882	6 by 8	16	13	86.7	473	148	84.2	742
	2 by 2	143	72.1	3,490	1,222	3,980	6 by 6	12	14	83.0	424	...	....	...
	....	...	....	....	....	....	6 by 7	9	13	74.7	477	...	....	...
	....	...	....	....	....	....	6 by 3	14	13	75.6	411	...	....	...
	....	...	....	....	....	....	6 by 2	12	12	66.5	430	...	....	...
	....	...	....	....	....	....	6 by 2	10	11	55.0	423	...	....	...
	....	...	....	....	....	....	6 by 2	8	12	56.7	396	...	....	...
	....	...	....	....	....	....	2 by 2	2	186	75.5	569			
Norway pine....	6 by 7	5	29.0	1,928	905	2,404	....	..	...	....	...	20	26.7	589
	4 by 7	8	28.4	2,154	1,063	2,652	....	..	...	....	...	...	....	...
	2 by 2	178	26.8	....	....	2,504								
Red spruce......	2 by 2	58	35.4	....	....	2,750	2 by 2	2	43	31.8	310	30	32.0	758
White spruce....	2 by 2	84	61.0	....	....	2.370	2 by 2	2	46	50.4	270	40	58.0	651

TABLE XX

BENDING TESTS ON AIR-SEASONED MATERIAL

Species	Sizes		Number of tests	Per cent of moisture	Rings per inch	F. S. at E.L.		M. of R.		M. of E.		Calculated shear	
	Cross section	Span				Average per square inch	Ratio to 2" by 2"	Average per square inch	Ratio to 2" by 2"	Average per square inch	Ratio to 2" by 2"	Average per square inch	Ratio to 2" by 2"
	Inches.	Ins.				Lbs.		Lbs.		1,000 lbs.		Lbs.	
Longleaf pine.....	8 by 16	180	5	22.2	16.0	3,390	0.50	4,274	0.37	1,747	1.00	288	0.75
	6 by 16	132	1	23.4	17.1	3,470	.51	6,610	.57	1,501	.86	388	1.01
	6 by 10	177	2	19.0	8.8	4,560	.68	7,880	.68	1,722	.99	214	.56
	4 by 11	180	1	18.4	23.9	3,078	.46	8,000	.69	1,660	.95	251	.66
	6 by 8	177	6	20.0	13.7	4,227	.63	8,196	.71	1,634	.94	177	.46
	2 by 2	30	17	15.9	13.9	6,750	1.00	11,520	1.00	1,740	1.00	383	1.00
Douglas fir.......	8 by 16	180	91	20.8	13.1	4,563	.68	6,372	.61	1,549	.91	269	.64
	5 by 8	180	30	14.9	12.2	5,065	.76	6,777	.65	1,853	1,09	218	.52
	2 by 2	24	211	19.0	16.4	6,686	1.00	10,378	1.00	1,695	1.00	419	1.00
Shortleaf pine.....	8 by 16	180	3	17.0	12.3	4,220	.54	6,030	.50	1,517	.85	398	.98
	8 by 14	180	3	16.0	12.3	4,253	.55	5,347	.44	1,757	.98	307	.76
	8 by 12	180	7	16.0	12.4	5,051	.65	7,331	.60	1,803	1 01	361	.89
	5 by 8	180	6	12.2	22.5	7,123	.92	9,373	.77	1,985	1.11	301	.74
	2 by 2	30	67	14.2	13.7	7,780	1.00	12,120	1 00	1,792	1.00	404	1.00
Western larch.....	8 by 16	180	23	18.3	21.9	3,343	.57	5,440	.53	1,409	.90	349	.96
	8 by 12	180	29	17.8	23.4	3,631	.62	6,186	.60	1,549	.99	295	.81
	5 by 8	180	10	13.6	27.6	4,730	.80	7,258	.71	1,620	1.04	221	.61
	2 by 2	30	240	16.1	26.8	5,880	1.00	10,254	1 00	1,564	1.00	364	1.00
Loblolly pine.....	8 by 16	180	14	20.5	7.4	4,195	.81	6,734	.72	1,619	1.10	462	1.45
	6 by 16	126	4	20.2	5.0	2,432	.47	4,295	.46	1,324	.90	266	.84
	6 by 10	174	3	21.3	4.7	3,100	.60	6,167	.66	1,449	.99	173	.54
	4 by 12	174	4	19.8	4.7	2,713	.52	5,745	.61	1,249	.85	185	.58
	8 by 8	180	9	22.9	4.9	2,903	.56	4,557	.48	1,136	.77	93	.29
	6 by 7	144	2	21.1	5.0	2,990	.58	4,968	.53	1,286	.88	116	.36
	4 by 8	132	8	19.5	9.1	3,384	.65	6,194	66	1,200	.82	196	.62
	2 by 2	30	123	17.6	6.6	5,170	1.00	9,400	1.00	1,467	1.00	318	1.00
Tamarack........	6 by 12	162	5	23.0	15.1	3,434	.45	5,640	.43	1,330	.82	318	.75
	4 by 10	162	4	14.4	9.7	4,100	.54	5,320	.41	1,356	.84	252	.59
	2 by 2	30	47	11.3	16.2	7,630	1.00	13,080	1.00	1,620	1.00	425	1.00
Western hemlock ..	8 by 16	180	44	17.7	17.8	4,398	.69	6,420	.62	1,737	1.04	406	1.06
	2 by 2	28	311	17.9	19.4	6,333	1.00	10,369	1.00	1,666	1.00	382	1.00
Redwood........	8 by 16	180	6	26.3	22.4	3,797	.79	4,428	.57	1,107	.96	294	1.05
	6 by 12	180	6	16.1	17.7	3,175	.66	3,353	.43	728	.64	167	.60
	7 by 9	180	6	15.9	15.2	3,280	.69	4,002	.51	1,104	.96	147	.53
	3 by 14	180	6	13.1	24.4	....	....	5,033	.64	....		291	1.04
	2 by 12	180	5	13.8	14.4	3,928	.82	5,336	.68	1,249	1.09	260	.93
	2 by 10	180	5	13.8	24.8	3,757	.79	4,606	.59	1,198	1.05	186	.67
	2 by 8	180	6	13.7	20.7	4,314	.90	5,050	.65	1,313	1.15	166	.60
	2 by 2	28	122	15.2	18.8	4,777	1.00	7,798	1.00	1,146	1.00	279	1.00
Norway pine......	6 by 12	162	5	16.7	8.1	2,968	.56	5,204	.61	1,123	.97	286	1.02
	4 by 10	162	5	13.7	12.0	5,170	.98	6,904	.82	1,712	1.48	317	1.13
	2 by 2	30	60	14.9	11.2	5,280	1.00	8,470	1.00	1,158	1.00	281	1.00

TABLE XXI

COMPRESSION AND SHEAR TESTS ON AIR-SEASONED MATERIAL

Species	Compression ∥ to grain						Compression ⊥ to grain					Shear		
	Size of specimen	Number of tests	Per cent of moisture	Cr. str. at E. L., per square inch	M. of E., per square inch	Cr. str. at max. ld., per square inch	Stress area	Height	Number of tests	Per cent of moisture	Cr. str. at E. L., per square inch	Number of tests	Per cent of moisture	Shear strength per square inch
	Inches			*Lbs.*	*1,000 lbs.*	*Lbs.*	*Inches*	*In.*			*Lbs.*			*Lbs.*
Longleaf pine....	4 by 5	46	26.3	3,480	....	4,800	4 by 5	4	22	25.1	572	52	20.2	984
Douglas fir......	6 by 6	259	20.3	3,271	1,038	4,258	4 by 8	16	44	20.8	732	465	22.1	822
	2 by 2	247	18.7	3,842	1,084	5,002	4 by 8	10	32	18.1	584	...	....	...
	.....	...	....	....	....	....	4 by 4	8	51	20.2	638	...	....	...
	.....	...	....	....	....	....	4 by 4	6	49	24.0	613	...	....	...
	.....	...	....	....	....	....	4 by 4	4	29	24.8	603	...	....	...
Shortleaf pine....	6 by 6	29	15.7	4,070	1,951	6,030	8 by 5	16	4	17.8	725	85	....	1,135
	2 by 2	57	14.2	....	....	6,380	8 by 5	14	3	16.3	757	...	....	...
	.....	...	....	....	....	....	8 by 5	12	5	15.1	730	...	....	...
	.....	...	....	....	....	....	5 by 5	8	6	13.0	918	...	....	...
	.....	...	....	....	....	....	2 by 2	2	57	13.9	926	...	....	...
Western larch...	6 by 6	112	16.0	....	....	5,445	8 by 6	16	17	18.8	491	193	15.0	905
	4 by 4	81	14.7	....	....	6,161	8 by 6	12	18	17.6	526	...	....	...
	2 by 2	270	14.8	....	....	5,934	5 by 4	8	22	13.3	735	...	....	...
Loblolly pine....	6 by 6	23	....	3 357	1,693	5,005	8 by 5	16	12	19.8	602	156	11.3	1,115
	5 by 5	10	22.4	2,217	545	2,950	8 by 5	8	7	22.9	679	...	....	...
	4 by 8	8	19.4	3,010	633	3,920	4 by 5	8	8	19.5	715	...	....	...
	2 by 2	69	....	....	....	5,547	...	..	...	....	....	...	....	...
Tamarack......	6 by 7	3	15.7	2,257	1,042	3,323	2 by 2	2	57	16.2	697	60	14.0	879
	4 by 7	3	13.6	3,780	1,301	4,823	.....	..	...	....	....	...	....	...
	4 by 4	57	14.9	3,386	1,353	4,346	.....	..	...	....	....	...	....	...
	2 by 2	66	14.6	....	....	4,790	.....	..	...	....	....	...	....	...
West. hemlock..	6 by 6	102	18.6	4,840	2,140	5,814	7 by 6	15	25	18.2	514	131	17.7	924
	2 by 2	463	17.0	4,560	1,923	5,403	6 by 6	6	26	16.8	431	...	....	...
	.....	...	....	....	....	....	4 by 4	4	6	15.9	488	...	....	...
Redwood.......	6 by 6	18	16.9	....	....	4,276	8 by 6	16	5	25.4	548	95	12.4	671
	2 by 2	115	14.6	....	....	5.119	6 by 6	12	6	14.7	610	...	....	...
	.....	...	....	....	....	....	7 by 6	9	5	14.8	500	...	....	...
	.....	...	....	....	....	....	3 by 6	14	2	12.6	470	...	....	...
	.....	...	....	....	....	....	2 by 6	12	2	16.2	498	...	....	...
	.....	...	....	....	....	....	2 by 6	10	4	14.3	511	...	....	...
	.....	...	....	....	....	....	2 by 6	8	2	13.2	429	...	....	...
	.....	...	....	....	....	....	2 by 2	2	145	13.8	564	...	....	...
Norway pine....	6 by 7	4	15.2	2,670	1,182	4,212	2 by 2	2	36	10.0	924	44	11.9	1,145
	4 by 7	2	22.2	3,275	1,724	4,575	.....	..	...	....	...	...	....	...
	4 by 4	55	16.6	3,048	1,367	4,217	.....	..	...	....	...	...	....	...
	2 by 2	44	11.2	....	....	7,550	.....	..	...	....	...	...	....	...

NOTE.—Following is an explanation of the abbreviations used in the foregoing tables:

F. S. at E. L. = Fiber stress at elastic limit.

M. of E. = Modulus of elasticity.

M. of R. = Modulus of rupture.

Cr. str. at E. L. = Crushing strength at elastic limit.

Cr. str. at max. ld. = Crushing strength at maximum load.

TABLE XXII

*WORKING UNIT-STRESSES FOR STRUCTURAL TIMBER †

EXPRESSED IN POUNDS PER SQUARE INCH

(From Manual of the American Railway Engineering Assn., 1911, p. 153)

NOTE.—The working unit-stresses given in this table are intended for railroad bridges and trestles. For highway bridges and trestles the unit-stresses may be increased twenty-five (25) per cent. For buildings and similar structures, in which the timber is protected from the weather and practically free from impact, the unit-stresses may be increased fifty (50) per cent. To compute the deflection of a beam under long-continued loading instead of that when the load is first applied, only fifty (50) per cent of the corresponding modulus of elasticity given in the table is to be employed.

Kind of timber	Ratio of length to depth	Formula for working-ing stress in long columns over 15 diameters	For columns under 15 diams., working stress	Compression parallel to the grain — Working stress	Compression parallel to the grain — Average ultimate	Compression perpendicular to the grain — Working stress	Compression perpendicular to the grain — Elastic limit	Longitudinal shear in beams — Working stress	Longitudinal shear in beams — Average ultimate	Parallel to the grain — Working stress	Parallel to the grain — Average ultimate	Modulus of elasticity — Average	Extreme fibre stress — Working stress	Extreme fibre stress — Average ultimate
Douglas fir	10	1200 (1−l/60d)	900	1200	3600	310	630	110	270	170	690	1,510,000	1200	6100
Longleaf pine	10	1300 (1−l/60d)	980	1300	3800	260	520	120	300	180	720	1,610,000	1300	6500
Shortleaf pine	10	1100 (1−l/60d)	830	1100	3400	170	340	130	330	170	710	1,480,000	1100	5600
White pine	10	1000 (1−l/60d)	750	1000	3000	150	290	70	180	100	400	1,130,000	900	4400
Spruce		1100 (1−l/60d)	830	1100	3200	130	370	70	170	150	600	1,310,000	1000	4800
Norway pine		800 (1−l/60d)	600	800	2600*	150	—	100	250	130	590*	1,190,000	800	4200
Tamarack		1000 (1−l/60d)	750	1000	3200*	220	—	100	260	170	670	1,220,000	900	4600
Western hemlock		1200 (1−l/60d)	900	1200	3500	220	440	100	270*	160	630	1,480,000	1100	5800
Redwood		900 (1−l/60d)	680	900	3300	150	400	—	—	80	300	800,000	900	5000
Bald cypress		1100 (1−l/60d)	830	1100	3900	170	340	—	—	120	500	1,150,000	900	4800
Red cedar		900 (1−l/60d)	680	900	2800	230	470	—	—	—	—	800,000	800	4200
White oak	12	1300 (1−l/60d)	980	1300	3500	450	920	110	270	210	840	1,150,000	1100	5700

l = Length in inches.
d = Least side in inches.

* Partially air-dry.

These unit-stresses are for a green condition of timber and are to be used without increasing the live load stresses for impact.

* Adopted, Vol. 1909, pp. 537, 564, 609–611.
† Green timber in exposed work.

BIBLIOGRAPHY

Part   I: Some general works on mechanics, materials of construction, and testing of materials.

Part II: Publications and articles on the mechanical properties
of wood, and timber testing.

Part III: Publications of the U. S. Government on the mechanical
properties of wood, and timber testing.

I. SOME GENERAL WORKS ON MECHANICS, MATERIALS OF CONSTRUCTION, AND TESTING OF MATERIALS

ALLAN, WILLIAM: Strength of beams under transverse loads. New York, 1893.

ANDERSON, SIR JOHN: The strength of materials and structures. London, 1902.

BARLOW, PETER: Strength of materials, 1st ed. 1817; rev. 1867.

BURR, WILLIAM H.: The elasticity and resistance of the materials of engineering. New York, 1911.

CHURCH, IRVING P.: Mechanics of engineering. New York, 1911.

HATFIELD, R. G.: Theory of transverse strain. 1877.

HATT, W. K., and SCOFIELD, H. H.: Laboratory manual of testing materials. New York, 1913.

JAMESON, J. M.: Exercises in mechanics. (Wiley technical series.) New York, 1913.

JAMIESON, ANDREW: Strength of materials. (Applied mechanics and mechanical engineering, Vol. II.) London, 1911.

JOHNSON, J. B.: The materials of construction. New York, 1910.

KENT, WILLIAM: The strength of materials. New York, 1890.

KOTTCAMP, J. P.: Exercises for the applied mechanics laboratory. (Wiley technical series.) New York, 1913.

LANZA, GAETANO: Applied mechanics. New York, 1901.

MERRIMAN, MANSFIELD: Mechanics of materials. New York, 1912.

MURDOCK, H. E.: Strength of materials. New York, 1911.

RANKINE, WILLIAM J. M.: A manual of applied mechanics. London, 1901.

THIL, A.: Conclusion de l'étude présentée à la Commission des méthodes d'essai des matériaux de construction. Paris, 1900.

THURSTON, ROBERT H.: A treatise on non-metallic materials of engineering: stone, timber, fuel, lubricants, etc. (Materials of engineering, Part I.) New York, 1899.

UNWIN, WILLIAM C.: The testing of materials of construction. London, 1899.

WATERBURY, L. A.: Laboratory manual for testing materials of construction. New York, 1912.

WOOD, DEVOLSON: A treatise on the resistance of materials. New York, 1897.

II. PUBLICATIONS AND ARTICLES ON THE MECHAN-ICAL PROPERTIES OF WOOD, AND TIMBER TESTING

ABBOT, ARTHUR V.: Testing machines, their history, construction and use. Van Nostrand's Eng. Mag., Vol. XXX, 1884, pp. 204–214; 325–344; 382–397; 477–490.

ADAMS, E. E.: Tests to determine the strength of bolted timber joints. Cal. Jour. of Technology, Sept., 1904.

ALVAREZ, ARTHUR C.: The strength of long seasoned Douglas fir and redwood. Univ. of Cal. Pub. in Eng., Vol. I, No. 2, Berkeley, 1913, pp. 11–20.

BARLOW, PETER: An essay on the strength and stress of timber. London, 1817; 3d ed., 1826.

————: Experiments on the strength of different kinds of wood made in the carriage department, Royal Arsenal, Woolwich. Jour. Franklin Inst., Vol. X, 1832, pp. 49–52. Reprinted from Philosophical Mag. and Annals of Philos., No. 63, Mch., 1832.

BATES, ONWARD: Pine stringers and floorbeams for bridges. Trans. Am. Soc. C. E., Vol. XXIII.

BAUSCHINGER, JOHANN: Untersuchungen über die Elasticität und Festigkeit von Fichten- und Kiefernbauhölzern. Mitt. a. d. mech.-tech. Laboratorium d. k. techn. Hochschule in München, 9. Hft., München, 1883.

————: Verhandlungen der Münchener Conferenz und der von ihr gewählten ständigen Commission zur Vereinbarung einheitlicher Prüfungsmethoden für Bau- und Constructions-material. Ibid., 14. Hft., 1886.

————: Untersuchungen über die Elasticität und Festigkeit verschiedener Nadelhölzer. Ibid., 16. Hft., 1887.

BEARE, T. HUDSON: Timber: its strength and how to test it. Engineering, London, Dec. 9, 1904.

BEAUVERIE, J.: Le bois. I. Paris, 1905, pp. 105–185.

————: Les bois industriels. Paris, 1910, pp. 55–77.

Bending tests with wood, executed at the Danish State Testing Laboratory, Copenhagen. Proc. Int. Assn. Test. Mat., 1912, XXIII₂, pp. 17. See also Eng. Record, Vol. LXVI, 1912, p. 269.

BERG, WALTER G.: Berg's complete timber test record. Chicago, 1899. Reprint from Am. Ry. Bridges and Buildings.

BOULGER, G. S.: Wood. London, 1908, pp. 112–121.

BOUNICEAU, —: Note et expériences sur la torsion des bois. [N.p., n.d.]

BOVEY, HENRY T.: Results of experiments at McGill University,

Montreal, on the strength of Canadian Douglas fir, red pine, white pine, and spruce. Trans. Can. Soc. C. E., Vol. IX, Part I, 1895, pp. 69–236.

BREUIL, M. PIERRE: Contribution to the discussion on the testing of wood. Proc. Int. Assn. Test. Mat., 1906, Disc. 1e, pp. 2.

BROWN, T. S.: An Account of some experiments made by order of Col. Totten, at Fort Adams, Newport, R. I., to ascertain the relative stiffness and strength of the following kinds of timber, *viz.*: white pine (*Pinus strobus*), spruce (*Abies nigra*), and southern pine (*Pinus australis*), also called long-leaved pine. Jour. Franklin Inst., Vol. VII (n. s.), 1831, pp. 230–238.

BUCHANAN, C. P.: Some tests of old timber. Eng. News, Vol. LXIV, No. 23, 1910, p. 67.

BUSGEN, M.: Zur Bestimmung der Holzhärten. Zeitschrift f. Forst- und Jagdwesen. Berlin, 1904, pp. 543–562.

CHEVANDIER, E., et WERTHEIM, G.: Mémoire sur les propriétés mécaniques du bois. Paris, 1846.

CIESLAR, A.: Studien über die Qualität rasch erwachsenen Fichtenholzes. Centralblatt f. d. ges. Forstwesen, Wien, 1902, pp. 337–403.

CLINE, McGARVEY: Forest Service investigations of American woods with special reference to investigations of mechanical properties. Proc. Int. Assn. Test. Mat., 1912, XXIII₅, pp. 17.

——————: Forest Service tests to determine the influence of different methods and rates of loading on the strength and stiffness of timber. Proc. Am. Soc. Test. Mat., Vol. VIII, 1908, pp. 535–540.

——————: The Forest Products Laboratory: its purpose and work. Proc. Am. Soc. Test. Mat., Vol. X, 1910, pp. 477–489.

——————: Specifications and grading rules for Douglas fir timber: an analysis of Forest Service tests on structural timbers. Proc. Am. Soc. Test. Mat., Vol. XI, 1911, pp. 744–766.

Comparative strength and resistance of various tie timbers. Elec. Traction Weekly, Chicago, June 15, 1912.

DAY, FRANK M.: Microscopic examination of timber with regard to its strength. 1883, pp. 6.

DEWELL, H. D.: Tests of some joints used in heavy timber framing. Eng. News, Mch. 19, 1914, pp. 594–598; *et seq.*

DÖRR, KARL: Die Festigkeit von Fichten- und Kiefernholz. Deutsche Bauzeitung, Berlin, Aug. 17, 1910. See also Zeitschrift d. ver. deutsch. Ing., Bd. 54, Nr. 36, 1910, p. 1503.

DUPIN, CHARLES: Expériences sur la flexibilité, la force, et l'élasticité des bois. Jour. de l'Ecole Polytechnique, Vol. X, 1815.

DUPONT, ADOLPHE, et BOUQUET DE LA GRYE: Les bois indigènes et étrangers. Paris, 1875, pp. 273–352.

ESTRADA, ESTEBAN DUQUE: On the strength and other properties of Cuban woods. Van Nostrand's Eng. Mag., Vol. XXIX, 1883, pp. 417–426; 443–449.

EVERETT, W. H.: Memorandum on mechanical tests of some Indian timbers. Govt. Bul. No. 6 (o.s.), Calcutta.

EXNER, WILHELM FRANZ: Die mechanische Technologie des Holzes. Wien, 1871. (A translation and revision of Chevandier and Wertheim's Mémoire sur les propriétés mécaniques du bois.)
————: Die technischen Eigenschaften der Hölzer. Lorey's Handbuch der Forstwissenschaft, II. Bd., 6. Kap., Tübingen, 1903.

FERNOW, B. E.: Scientific timber testing. Digest of Physical Tests, Vol. I, No. 2, 1896, pp. 87–95.

FOWKE, FRANCIS: Experiments on British colonial and other woods. 1867.

GARDNER, ROLAND: I. Mechanical tests, properties, and uses of thirty Philippine woods. II. Philippine sawmills, lumber market and prices. Bul. 4, Bu. For., P. I., 1906. (2d ed., 1907, contains tests of 34 woods.)

GAYER, KARL: Forest utilization. (Vol. V, Schlich's Manual of Forestry. Translation of Die Forstbenutzung, Berlin, 1894.) London, 1908.

GOLLNER, H.: Ueber die Festigkeit des Schwarzföhrenholzes. Mitt. a. d. forstl. Versuchswesen Oesterreichs. II. Bd., 3. Hft., Wien, 1881.

GOTTGETREU, RUDOLPH: Physische und chemische Beschaffenheit der Baumaterialien. 3d ed., Berlin, 1880.

GREEN, A. O.: Tasmanian timbers: their qualities and uses. Hobart, Tasmania, 1903, pp. 63.

GREGORY, W. B.: Tests of creosoted timber. Trans. Am. Soc. C. E., Vol. LXXVI, 1913, pp. 1192–1203. See also ibid., Vol. LXX, p. 37.

GRISARD, JULES, et VANDENBERGHE, MAXIMILIEN: Les bois industriels, indigènes et exotiques; synonymie et description des espèces, propriétés physiques des bois, qualités, défauts, usages et emplois. Paris, 189–. From Bul. de la Société nationale d'acclimatation de France, Vols. XXXVIII–XL.

Hardwoods of Western Australia. Engineering, Vol. LXXXIII, Jan. 11, 1907, pp. 35–37.

HATT, WILLIAM KENDRICK: A Preliminary program for the timber test work to be undertaken by the Bureau of Forestry, United States Department of Agriculture. Proc. Am. Soc. Test. Mat., Vol. III, 1903, pp. 308–343. Appendix I: Method of determining the effect of the rate of application of load on the strength of timber, pp. 325–327; App. II: A discussion on the effect of moisture on strength and stiffness of timber, together with a plan of procedure for future tests, pp. 328–334.

HATT, WILLIAM KENDRICK: Relation of timber tests to forest products. Proc. Int. Assn. Test. Mat., 1906, C 2 e, pp. 6.

————: Structural timber. Proc. Western Ry. Club, St. Louis, Mch. 17, 1908.

————: Abstract of report on the present status of timber tests in the Forest Service, United States Department of Agriculture. Proc. Int. Assn. Test. Mat., 1909, XVI₁, pp. 10.

———— and TURNER, W. P.: The Purdue University impact machine. Proc. Am. Soc. Test. Mat., Vol. VI, 1906, pp. 462–475.

HAUPT, HERMAN: Formula for the strain upon timber. Center of gravity of an ungula and semi-cylinder. Jour. Franklin Inst., Vol. XIX, 3d series, 1850, pp. 408–413.

HEARDING, W. H.: Report upon experiments . . . upon the compressive power of pine and hemlock timber. Washington, 1872, pp. 12.

HOWE, MALVERD A.: Wood in compression; bearing values for inclined cuts. Eng. News, Vol. LXVIII, 1912, pp. 190–191.

HOYER, EGBERT: Lehrbuch der vergleichenden mechanischen Technologie. 1878.

IHLSENG, MANGUS C.: On the modulus of elasticity of some American woods as determined by vibration. Van Nostrand's Eng. Mag., Vol. XIX, 1878, pp. 8–9.

————: On a mode of measuring the velocity of sounds in woods. Am. Jour. Sci. and Arts, Vol. XVII, 1879.

JACCARD, P.: Etude anatomique des bois comprimés. Mitt. d. Schw. Centralanstalt f. d. forst. Versuchswesen. X. Bd., 1. Hft., Zurich, 1910, pp. 53–101.

JANKA, GABRIEL: Untersuchungen über die Elasticität und Festigkeit der österreichischen Bauhölzer. I. Fichte Südtirols; II. Fichte von Nordtirol vom Wienerwalde und Erzgebirge; III. Fichte aus den Karpaten, aus dem Böhmerwalde, Ternovanerwalde und den Zentralalpen. Technische Qualität des Fichtenholzes im allgemeinen; IV. Lärche aus dem Wienerwalde, aus Schlesien, Nord- und Südtirol. Mitt. a. d. forst. Untersuchungswesen Oesterreichs, Wien, 1900–13.

————: Untersuchungen über Holzqualität. Centralblatt f. d. ges. Forstwesen. Wien, 1904, pp. 95–115.

————: Ueber neuere holztechnologische Untersuchungen. Oesterr. Vierteljahresschrift für Forstwesen, Wien, 1906, pp. 248–269.

————: Die Härte des Holzes. Centralblatt f. d. ges. Forstwesen, Wien, 1906, pp. 193–202; 241–260.

Janka, Gabriel: Die Einwirkung von Süss- und Salzwässern auf die gewerblichen Eigenschaften der Hauptholzarten. I. Teil. Untersuchungen u. Ergebnisse in mechanisch - technischer Hinsicht. Mitt. a. d. forst. Versuchswesen Oesterreichs, 33. Hft., Wien, 1907.

————: Results of trials with timber carried out at the Austrian forestry testing-station at Mariabrunn. Proc. Int. Assn. Test. Mat., 1906, Disc. 2 e, pp. 7.

————: Ueber die an der k. k. forstlichen Versuchsanstalt Mariabrunnen gewonnenen Resultate der Holzfestigkeitsprüfungen. Zeitschrift d. Oesterr. Ing. u. Arch. Ver., Wien, Aug. 9, 1907.

————: Ueber Holzhärteprüfung. Centralblatt f. d. ges. Forstwesen, Wien, 1908, pp. 443–456.

————: Testing the hardness of wood by means of the ball test. Proc. Int. Assn. Test. Mat., 1912, XXIII₃.

Jenny, K.: Untersuchungen über die Festigkeit der Hölzer aus den Ländern der ungarischen Krone. Budapest, 1873.

Johnson, J. B.: Time tests of timber in endwise compression. Paper before Section D, Am. Assn. for Adv. of Sci., Aug., 1898.

Johnson, Walter B.: Experiments on the adhesion of iron spikes of various forms when driven into different species of timbers. Jour. Franklin Inst., Vol. XIX (n. s.), 1837, pp. 281–292.

Julius, G. A.: Western Australia timber tests, 1906. The physical characteristics of the hardwoods of Western Australia. Perth, 1906, pp. 36.

————: Supplement to the Western Australia timber tests, 1906. The hardwoods of Australia. Perth, 1907, pp. 6.

Karmarsh, Carl: Handbuch der mechanischen Technologie. I. Aufl., 1837; V. Aufl., 1875; verm. von H. Fisher, 1888.

Kidder, F. E.: Experiments on the transverse strength of southern and white pine. Van Nostrand's Eng. Mag., Vol. XXII, 1880, pp. 166–168.

————: Experiments on the strength and stiffness of small spruce beams. Ibid., Vol. XXIV, 1881, pp. 473–477.

————: Experiments on the fatigue of small spruce beams. Jour. Franklin Inst., Vol. CXIV, 1882, pp. 261–279.

Kidwell, Edgar: The efficiency of built-up wooden beams. Trans. Am. Inst. Min. Eng., Feb., June, 1898.

Kirkaldy, Wm. G.: Illustrations of David Kirkaldy's system of mechanical testing. London, 1891.

Kummer, Frederick A.: The effects of preservative treatment on the strength of timber. Proc. Am. Soc. Test. Mat., Vol. IV, 1904, pp. 434–438.

LABORDÈRE, P., and ANSTETT, F.: Contribution to the study of means for improving the strength of wood for pavements. Proc. Int. Assn. Test. Mat., 1912, XXIII, pp. 12.

LANZA, GAETANO: An account of certain tests on the transverse strength and stiffness of large spruce beams. Trans. Am. Soc. Mech. Eng., Vol. IV, 1882, pp. 119–135. See also Jour. Franklin Inst., Vol. XCV, 1883, pp. 81–94.

LASLETT, T.: Properties and characteristics of timber. Chatham, 1867.

————: Timber and timber trees, native and foreign. (2d ed. revised and enlarged by H. Marshall Ward.) London and New York, 1894.

LEA, W.: Tables of strength and deflection of timber. London, 1861.

LEDEBUR, A.: Die Verarbeitung des Holzes auf mechanischem Wege. 1881.

LORENZ, N. VON: Analytische Untersuchung des Begriffes der Holzhärte. Centralblatt f. d. ges. Forstwesen, Wien, 1909, pp. 348–387.

LUDWIG, PAUL: Die Regelprobe. Ein neues Verfahren zur Härtebestimmung von Materialien. Berlin. 1908.

MACFARLAND, H. B.: Tests of longleaf pine bridge timbers. Bul. 149, Am. Ry. Eng. Assn., Sept., 1912. See also Eng. News, Dec. 12, 1912, p. 1035.

McKAY, DONALD: On the weight and strength of American shiptimber. Jour. Franklin Inst., Vol. XXXIX (3d series), 1860, p. 322.

MALETTE, J.: Essais des bois de construction. Revue Technique, Apr. 25, 1905.

MANN, JAMES: Australian timber: its strength, durability, and identification. Melbourne, 1900.

MARTIN, CLARENCE A.: Tests on the relation between crossbending and direct compressive strength in timber. Railroad Gazette, Mch. 13, 1903.

Methods of testing metals and alloys . . . Recommended by the Fourth Congress of the International Association for Testing Materials, held at Brussels, Sept. 3–6, 1906. London, 1907, pp. 54. Methods of testing wood, pp. 39–49.

MIKOLASCHEK, CARL: Untersuchungen über die Elasticität und Festigkeit der wichtigsten Bau- und Nutzhölzer. Mitt. a. d. forstl. Versuchswesen Oesterreiches, II. Bd., 1. Hft., Wien, 1879.

MOELLER, JOSEPH: Die Rohstoffe des Tischler- und Drechslergewerbes. I. Theil: Das Holz. Kassel, 1883, pp. 68–122.

MOLESWORTH, G. L.: Graphic diagrams of strength of teak beams. Roorke, 1881.

Morgan, J. J.: Bending strength of yellow pine timber. Eng. Record, Vol. LXVII, 1913, pp. 608–609.

Moroto, K.: Untersuchungen über die Biegungselasticität und -Festigkeit der japanischen Bauhölzer. Centralblatt f. d. ges. Forstwesen, Wien, 1908, pp. 346–355.

Nördlinger, H.: Die technischen Eigenschaften der Hölzer für Forst- und Baubeamte, Technologen und Gewerbetreibende. Stuttgart, 1860.

——————: Druckfestigkeit des Holzes. 1882.

——————: Die gewerblichen Eigenschaften der Hölzer. Stuttgart, 1890.

North, A. T.: The grading of timber on the strength basis. Address before Western Society of Engineers. Lumber World Review, May 25, 1914, pp. 27–29.

Norton, W. A.: Results of experiments on the set of bars of wood, iron, and steel, after a transverse stress. Van Nostrand's Eng. Mag., Vol. XVII, 1877, pp. 531–535.

Paccinotti e Peri: [Investigations into the elasticity of timbers.] Il Cimento, Vol. LVIII, 1845.

Palacio, E.: Tensile tests of timber. La Ingenieria, Buenos Aires, May 31, 1903, et seq.

Parent, —: Expériences sur la résistance des bois de chêne et de sapin. Mémoires de l'Académie des Sciences, 1707–08.

Propositions relatives à l'établissement d'un procédé uniforme pour l'essai des qualités techniques des bois. Proc. Int. Assn. Test. Mat., 1901, Annexe, pp. 13–28.

Rogers, Charles G.: A manual of forest engineering for India. Vol. I, Calcutta, 1900, pp. 50–91.

Rudeloff, M.: Der heutige Stand der Holzuntersuchungen. Mitt. a. d. königlichen tech. Versuchsanstalt, Berlin, IV, 1899.

——————: Principles of a standard method of testing wood. Proc. Int. Assn. Test. Mat., 1906, 23 C, pp. 16.

——————: Large vs. small test-pieces in testing wood. Proc. Int. Soc. Test. Mat., 1912, XXIII₁, pp. 7.

Sargent, Charles Sprague: Woods of the United States, with an account of their structure, qualities, and uses. New York, 1885.

Schneider, A.: Zusammengesetzte Träger. Zeitschrift d. Oesterr. Ing. u. Arch. Ver., Nov. 24; Dec. 9, 1899.

Schwappach, A. F.: Beiträge zur Kenntniss der Qualität des Rotbuchenholzes. Zeitschrift f. Forst- und Jagdwesen, Berlin, 1894, pp. 513–539.

——————: Untersuchungen über Raumgewicht und Druckfestigkeit des Holzes wichtiger Waldbäume. Berlin, 1897–98.

——————: Etablissement de méthodes uniformes pour l'essai

à la compression des bois.   Proc. Int. Assn. Test. Mat., 1901, Rapport 23, pp. 28.

SEBERT, H.: Notice sur les bois de la Nouvelle Calédonie suivie de considérations générales sur les propriétés mécaniques des bois et sur les procédés employés pour les mesurer. Paris.

SHERMAN, EDWARD C.: Crushing tests on water-soaked timbers. Eng. News, Vol. LXII, 1909, p. 22.

SNOW, CHARLES H.: The principal species of wood: their characteristic properties. New York, 1908.

STAUFFER, OTTMAR: Untersuchungen über specifisches Trockengewicht, sowie anatomisches Verhalten des Holzes der Birke. München, 1892.

STENS, D.: Ueber die Eigenschaften imprägnierter Grubenhölzer, insbesondere über ihre Festigkeit. Glückauf, Essen, Mch. 6, 1907.

Strength of wood for pavements. Can. Eng., Toronto, Sept. 12, 1912.

STÜBSCHEN-KISCHNER: Karmarsch-Heerins technisches Wörterbuch. 3. Aufl., 1886.

TALBOT, ARTHUR N.: Tests of timber beams. Bul. 41, Eng. Exp. Sta., Univ. of Ill., Urbana, 1910.

Tests of wooden beams made at the Massachusetts Institute of Technology on spruce, white pine, yellow pine, and oak beams of commercial sizes. Technology Quarterly, Boston, Vol. VII, 1894.

TETMAJER, L. v.: Zur Frage der Knickungsfestigkeit der Bauhölzer. Schweizerische Bauzeitung, Bd. 11, Nr. 17.

————: Methoden und Resultate der Prüfung der schweizerischen Bauhölzer. Mitt. d. Anstalt z. Prüfung v. Baumaterialien am eidgenössischen Polytechnicum in Zürich. 2. Hft., 1884.

————: Methoden und Resultate der Prüfung der schweizerischen Bauhölzer. Mitt. d. Materialprüfungs-Anstalt am Schweiz. Polytechnikum in Zürich. Landesaustellungs-Ausgabe, 2. Hft., Zürich, 1896.

THELEN, ROLF: The structural timbers of the Pacific Coast. Proc. Am. Soc. Test. Mat., Vol. VIII, 1908, pp. 558–567.

THURSTON, R. H.: Torsional resistance of materials determined by a new apparatus with automatic registry. Jour. Franklin Inst., Vol. LXV, 1873, pp. 254–260.

————: On the strength of American timber. Ibid., Vol. LXXVIII, 1879, pp. 217–235.

————: Experiments on the strength of yellow pine. Ibid., Vol. LXXIX, 1880, pp. 157–163.

————: Influence of time on bending strength and elasticity. Proc. Am. Assn. for Adv. Sci., 1881. Also Proc. Inst. C. E., Vol. LXXI.

————: On the effect of prolonged stress upon the strength

and elasticity of pine timber. Jour. Franklin Inst., Vol. LXXX, 1881, pp. 161–169.

THURSTON, R. H.: On Flint's investigations of Nicaraguan woods. *Ibid.*, Vol. XCIV, 1887, pp. 289–315.

TIEMANN, HARRY DONALD: The effect of moisture and other extrinsic factors upon the strength of wood. Proc. Am. Soc. Test. Mat., Vol. VII, 1907, pp. 582–594.

————: The effect of the speed of testing upon the strength of wood and the standardization of tests for speed. *Ibid.*, Vol. VIII, 1908, pp. 541–557.

————: The theory of impact and its application to testing materials. Jour. Franklin Inst., Vol. CLXVIII, 1909, pp. 235–259; 336–364.

————: Some results of dead load bending tests of timber by means of a recording deflectometer. Proc. Am. Soc. Test. Mat., Vol. IX, 1909, pp. 534–548.

TJADEN, M. E. H.: Het Indrukken van Paalkoppen in Kespen. De Ingenieur, Sept. 11, 1909.

————: Weerstand van Hout loodrecht op de Vezelrichting. *Ibid.*, May, 1911.

————: Buigrastheid van Hout. *Ibid.*, May 31, 1913.

TRAUTWINE, JOHN C.: Shearing strength of some American woods. Jour. Franklin Inst., Vol. CIX, 1880, pp. 105–106.

TREDGOLD, THOMAS: Elementary principles of carpentry. London, 1870.

TURNBULL, W.: A practical treatise on the strength and stiffness of timber. London, 1833.

Untersuchungen über den Einfluss des Blauwerdens auf die Festigkeit von Kiefernholz. Mitt. a. d. könig. techn. Versuchsanstalten, I, 1897.

Verfahren zur Prüfung v. Metallen und Legierungen, von hydraulischen Bindemitteln, von Holz, von Ton-, Steinzeug- und Zementröhren. Empfohlen v. dem in Brüssel v. 3–6, IX, 1906, abgeh. IV. Kongress des internationalen Verbandes f. die Materialprüfungen der Technik. Wien, 1907.

WARREN, W. H.: Australian timbers. Sydney, 1892.

————: The strength, elasticity, and other properties of New South Wales hardwood timbers. Sydney, 1911.

————: The strength, elasticity, and other properties of New South Wales hardwood timbers. Proc. Int. Assn. Test. Mat., 1912, XXIII$_6$, pp. 9.

————: The properties of New South Wales hardwood timbers. Builder, London, Nov. 1, 1912.

————: The hardwood timbers of New South Wales, Australia. Jour. Soc. of Arts, London, Dec. 6, 1912.

WELLINGTON, A. M.: Experiments on impregnated timber. Railroad Gazette, 1880.

WIJKANDER, ——: Untersuchung der Festigkeitseigenschaften schwedischer Holzarten in der Materialprüfungsanstalt des Chalmers'schen Institutes ausgeführt. 1897.

WING, CHARLES B.: Transverse strength of the Douglas fir. Eng. News, Vol. XXXIII, Mch. 14, 1895.

III. PUBLICATIONS OF THE U. S. GOVERNMENT ON THE MECHANICAL PROPERTIES OF WOOD, AND TIMBER TESTING

MISCELLANEOUS

House Misc. Doc. 42, pt. 9, 47th Cong., 2d sess., 1884. (Vol. IX, Tenth Census report.) Report on the forests of North America (exclusive of Mexico). Part II, The Woods of the United States.

House Report No. 1442, 53d Cong., 2d sess. Investigations and tests of American timber. 1894, pp. 4.

War Dept. Doc. 1. Resolutions of the conventions held at Munich, Dresden, Berlin, and Vienna, for the purpose of adopting uniform methods for testing construction materials with regard to their mechanical properties. By J. Bauschinger. Translated by O. M. Carter and E. A. Gieseler. 1896, pp. 44.

War Dept. Doc. 11. On tests of construction materials. Translations from the French and from the German. By O. M. Carter and E. A. Gieseler. 1896, pp. 84.

House Doc. No. 181, 55th Cong., 3d sess. Report upon the forestry investigations of the U. S. Department of Agriculture, 1877–1898. By B. E. Fernow, 1899, pp. 401. Contains chapter on The work in timber physics in the Division of Forestry, by Filibert Roth, pp. 330–395.

FOREST SERVICE

Cir. 7—The Government timber tests [189–], pp. 4.

Cir. 8—Strength of "boxed" or "turpentine" timber. 1892, pp. 4.

Bul. 6—Timber Physics. Pt. I. Preliminary report. 1. Need of the investigation. 2. Scope and historical development of the science of "timber physics." 3. Organization and methods of timber examinations in the Division of Forestry. By B. E. Fernow, 1892, pp. 57.

Unnumbered Cir.—Instructions for the collection of test pieces of pines for timber investigations [1893], pp. 4.

Cir. 9—Effect of turpentine gathering on the timber of long-leaf pine. By B. E. Fernow [1893], p. 1.

Bul. 8—Timber physics. Pt. II. Progress report. Results of investigations on longleaf pine. 1893, pp. 92.

Bul. 10—Timber: an elementary discussion of the characteristics and properties of wood. By Filibert Roth. 1895, pp. 88.

Bul. 12—Economical designing of timber trestle bridges. By A. L. Johnson, 1896, pp. 57.

Cir. 12—Southern pine, mechanical and physical properties. 1896, pp. 12.

Cir. 15—Summary of mechanical tests on thirty-two species of American woods. 1897, pp. 12.

Cir. 18—Progress in timber physics. 1898, pp. 20.

Cir. 19—Progress in timber physics: Bald cypress (*Taxodium distichum*). By Filibert Roth, 1898, pp. 24.

Y. B. Extr. 288—Tests on the physical properties of woods. By F. E. Olmstead, 1902, pp. 533–538.

Unnumbered Cir.—Timber tests. [1903], pp. 15.

Unnumbered Cir.—Timber preservation and timber testing at the Louisiana Purchase Exposition. 1904, pp. 6.

Cir. 32—Progress report on the strength of structural timber. By W. K. Hatt, 1904, pp. 28.

Bul. 58—The red gum. By Alfred Chittenden. Includes a discussion of The mechanical properties of red gum wood, by W. K. Hatt. 1905, pp. 56.

Cir. 38—Instructions to engineers of timber tests. By W. K. Hatt, 1906, pp. 55. Revised edition, 1909, pp. 56.

Cir. 39—Experiments on the strength of treated timber. By W. K. Hatt, 1906, pp. 31. Revised edition, 1908.

Bul. 70—Effect of moisture upon the strength and stiffness of wood. By H. D. Tiemann, 1906, pp. 144.

Cir. 46—Holding force of railroad spikes in wooden ties. By W. K. Hatt, 1906, pp. 7.

Cir. 47—Strength of packing boxes of various woods. By W. K. Hatt, 1906, pp. 7.

Cir. 108—The strength of wood as influenced by moisture. By H. D. Tiemann, 1907, pp. 42.

Cir. 115—Second progress report on the strength of structural timber. By W. K. Hatt, 1907, pp. 39.

Cir. 142—Tests of vehicle and implement woods. By H. B. Holroyd and H. S. Betts, 1908, pp. 29.

Cir. 146—Experiments with railway cross-ties. By H. B. Eastman, 1908, pp. 32.

Cir. 179—Utilization of California eucalypts. By H. S. Betts and C. Stowell Smith, 1910, pp. 30.

Bul. 75—California tanbark oak.   Part II, Utilization of the wood of tanbark oak, by H. S. Betts, 1911, pp. 24–32.

Bul. 88—Properties and uses of Douglas fir.   By McGarvey Cline and J. B. Knapp, 1911, pp. 75.

Cir. 189—Strength values for structural timbers.   By McGarvey Cline, 1912, pp. 8.

Cir. 193—Mechanical properties of redwood.   By A. L. Heim, 1912, pp. 32.

Bul. 108—Tests of structural timbers.   By McGarvey Cline and A. L. Heim, 1912, pp. 1231.

Bul. 112—Fire-killed Douglas fir: a study of its rate of deterioration, usability, and strength.   By J. B. Knapp, 1912, pp. 18.

Bul. 115—Mechanical properties of western hemlock.   By O. P. M. Goss, 1913, pp. 45.

Bul. 122—Mechanical properties of western larch.   By O. P. M. Goss, 1913, pp. 45.

Cir. 213—Mechanical properties of woods grown in the United States.   1913, pp. 4.

Cir. 214—Tests of packing boxes of various forms.   By John A. Newlin, 1913, pp. 23.

Review Forest Service Investigations.   1913.   [Outline of investigations.]   Vol. I, pp. 17–21.   A microscopic study of the mechanical failure of wood, by Warren D. Brush.   Vol. II, pp. 33–38.

Bul. 67, U. S. D. A.—Tests of Rocky Mountain woods for telephone poles.   By Norman deW. Betts and A. L. Heim, 1914, pp. 28.

Bul. 77, U. S. D. A.—Rocky Mountain mine timbers.   By Norman deW. Betts, 1914, pp. 34.

Bul. 86, U. S. D. A.—Tests of wooden barrels.   By J. A. Newlin, 1914, pp. 12.

REPORTS OF TESTS ON THE STRENGTH OF STRUCTURAL MATERIAL, MADE AT THE WATERTOWN ARSENAL, MASS.

House Ex. Doc. No. 12, 47th Cong., 1st sess., 1882.   Strength of wood grown on the Pacific slope, pp. 19–93.

Senate Ex. Doc. No. 1, 47th Cong., 2d sess., 1883.   Resistance of white and yellow pines to forces of compression in the direction of the fibers, as used for columns, or posts, pp. 239–395.

Senate Ex. Doc. No. 5, 48th Cong., 1st sess., 1884.   Tests of California laurel wood by compression, indentation, shearing, transverse tension, pp. 223–236.   Tests of North American woods (under supervision of Prof. C. S. Sargent in charge of the forestry division of the Tenth Census), with 16 photographs of fractures of American woods, pp. 237–347.

Senate Ex. Doc. No. 35, 49th Cong., 1st sess., 1885. Adhesion of nails, spikes, and screws in various woods. Experiments on the resistance of cut nails, wire nails (steel), wood screws, lag screws in white pine, yellow pine, chestnut, white oak, and laurel, pp. 448–471.

House Ex. Doc. No. 14, 51st Cong., 1st sess., 1890. Adhesion of spikes and bolts in railroad ties, pp. 595–617.

House Ex. Doc. No. 161, 52d Cong., 1st sess., 1892. Adhesion of nails in wood, pp. 744–745.

House Ex. Doc. No. 92, 53d Cong., 3d sess., 1895. Woods—compression tests (endwise compression), pp. 471–476.

House Doc. No. 54, 54th Cong., 1st sess., 1896. Compression tests on Douglas fir wood, pp. 536–563. Expansion and contraction of oak and pine wood, pp. 567–574.

House Doc. No. 164, 55th Cong., 2d sess., 1898. Compression tests of timber posts, pp. 405–411. New posts of yellow pine and spruce, pp. 413–450; Old yellow pine posts from Boston Fire Brick Co. building, No. 394 Federal St., Boston, Mass., pp. 451–473.

House Doc. No. 143, 55th Cong., 3d sess., 1899. Fire-proofed wood (endwise and transverse tests), pp. 676–681.

House Doc. No. 190, 56th Cong., 2d sess., 1901. Cypress wood for United States Engineer Corps; compression and transverse tests, pp. 1121–1126. Old white pine and red oak from roof trusses of Old South Church, Boston, Mass., pp. 1127–1130. Compression of rubber, balata, and wood buffers, pp. 1149–1158.

House Doc. No. 335, 57th Cong., 2d sess., 1903. Douglas fir and white oak woods. Transverse and shearing tests; also observations on heat conductivity of sticks over wood fires and a stick exposed to low temperature. Expansion crosswise the grain of wood after submersion, pp. 519–561. Adhesion of lag screws and bolts in wood, pp. 563–578.

INDEX

PAGE

ABRASION............39, 114–117
Annual rings................ 44
Apparatus, testing, 94, 99, 102, 104,
   107, 110–111, 114, 117, 118,
     121, 122, 132, 133, 136
Arborvitæ, 6, 9, 13, 16, 20, 27, 32,
               42, 57
Ash........15, 22, 44, 48, 51, 66, 78
   black......20, 27, 32, 40, 42, 56
   white, 9, 13, 16, 20, 27, 32, 40,
            42, 45, 56
Aspen, largetooth............ 13
Axis, neutral................ 23

BASSWOOD, 9, 13, 16, 20, 27, 32, 40
           42, 44, 56
Beams............15, 24–37, 92, 94
   cantilever................ 24
   continuous............... 24
   simple................... 24
Beech, 9, 13, 16, 20, 22, 27, 32, 40,
         42, 51, 57
Bending large beams.........94–99
   small beams.......99–102, 132
   strength.......2, 22–37, 26, 75
Bibliography.............145–160
Birch....................22, 44
   yellow, 9, 13, 16, 20, 27, 32, 40,
          42, 57
Bird-peck.................59, 72
Black check................. 59
Boiling, effect of............6, 85
Bow, flexure of a............ 4
Boxheart.................54, 82
Brash...................... 6
Breaking strength of beams.... 15
Brittleness...........6, 34, 37, 38
Buckeye...................13, 44
Buckling of fibres...........15, 77
Butternut.................. 13

CANTILEVER................ 24
Calibration of testing machines, 92,112
Case-hardening.............. 80
Catalpa.................... 44
Cedar, Central American....... 22
   incense......13, 16, 20, 27, 57
   red.................... 144
   white.................. 22
Checking.........54, 61, 74, 75–84
Chelura.................... 67
Cherry, black.............13, 22
Chestnut, 15, 22, 44, 49, 50, 51, 66, 78
Cleavability.................2, 41
Cleavage.............41, 118, 133
Coefficient of elasticity (see Mod-
   ulus of elasticity)
   expansion............... 84
Cold, effect of.. 86
Color..................50, 58–59
Column, long..........12, 14, 144
   short.............15, 102–104
Compression across the grain,
         94, 104–107, 133
   endwise.....12, 92, 94, 102, 132
   failure.................34, 104
   parallel to grain (see C. end-
     wise)
   perpendicular to grain (see
     C. right angles to grain)
   right angles to grain....94, 133
Compressive strength.......1, 9, 23
Compressometer............. 103
Coniferous wood............. 44
Cottonwood..............44, 55
Creosote, effect of........... 87
Cross-arms, testing........... 124
Cross grain...............8, 59–61
Cross-grained tension failure.... 34
Crushing strength.......2, 9, 54, 75
   formula for............. 104

PAGE

Cucumber tree............... 13
Cup shake.................. 64
Cypress, bald, 9, 13, 16, 20, 27, 32, 40,
42, 57, 68, 144

DEAD LOAD (see Load)
Definitions.................. 2–7
Deflection..................25, 30
    measuring..........96–97, 100
Deflectometer..........99, 106–107
Deformation, measuring....103, 107
Density....................54–58
Diffuse-porous............44, 50
Dogwood................... 22
Drying....................75–84
    effect of........75–78, 138–139
Dry rot..................... 69
Durability..............53, 54, 75

EARLY WOOD..............44, 82
Ease of working, factors affecting,
48, 50
Ebony..................... 22
Elasticity, modulus of.....6, 25, 89
    formulæ for, 98,102, 104, 114, 123
Elastic limit...........2, 5, 22, 62
    resilience............... 6
    formulæ for......98, 102, 104
Elm....................8, 38, 44
    rock, 9, 13, 16, 20, 27, 32, 40, 42,
56
    slippery.......13, 27, 32, 40, 56
    white...13, 20, 27, 32, 40, 42, 56
Elongation................3, 7, 33
Eucalyptus globulus.......54, 78, 82
    viminalis................ 54

FACTOR OF SAFETY.........29, 139
Failures, bending......33–37, 77, 78
    compression, endwise, 12, 15–19,
104
    cross-grained....10, 76, 77, 107
    shearing................. 19
    tension................. 8
    torsion..................38, 39

PAGE

Fibre-saturation point........ 78
Fibre strain, rate of.........92–93
    stress................... 15
    at elastic limit.54, 62, 75, 123
    formulæ, 98, 102, 104, 107,
114
Fir, Alpine, 16, 20, 27, 32, 40, 42, 57
    amabilis.........16, 20, 27, 57
    Douglas, 13, 16, 20, 27, 32, 36,
40, 42, 48, 55, 57, 124, 140,
141, 142, 143, 144
    white, 9, 16, 20, 27, 32, 40, 42, 57
Flexibility...................5, 37
Flexure................. 4, 12, 60
Formulæ, 98, 102, 104, 107, 110, 114,
123
Frost splits.................62–64
Fungi...............59, 68–70, 75

GRAIN, cross..............8, 59–61
    diagonal................ 60
    spiral.................. 60
Growth, in diameter........43, 44
    locality of, effect........70–73
    rate of, effect.......43,–50, 72
    rings..................44, 52
        measuring..........95–96
Gum..................22, 44, 51
    red.................27, 45, 56

HACKBERRY, 9, 13, 16, 20, 27, 32, 42,
51, 56
Hardness, 2, 39–41, 54, 114–118, 133
Heart break................. 65
    shake.................. 64
Heartwood........50–54, 58, 73, 75
Heat, effect of..............84–86
Hemlock, 9, 13, 15, 16, 20, 22, 27, 32,
40, 42, 44, 45, 56, 78
    western, 48, 59, 140, 141, 142,
143, 144
Hickory, 8, 22, 38, 40, 43, 44, 49, 51,
53, 55, 59, 65, 66, 72, 124
    big shellbark..13, 16, 20, 27, 56
    bitternut.....13, 16, 20, 27, 56

PAGE

Hickory, mockernut.13, 16, 20, 27, 56
    nutmeg.......13, 16, 20, 27, 56
    pignut, 6, 7, 13, 16, 20, 27, 48, 56
    shagbark.....13, 16, 20, 27, 56
    water........13, 16, 20, 27, 56
Hollow-horning.............. 80
Honey-combing.............54, 80

Impact........30–33, 110–114, 132
Implement woods, testing...... 124
Indentation...........39, 117–118
Injuries, fungous...... 52, 59, 68–70
    insect...........52, 66–67, 72
    marine wood-borer.......67–68
    parasitic plant............ 70

Kerosene, effect of........... 87
Knots............51, 52, 61–62, 89

Larch..................... 8
    western, 36, 48, 64, 140, 141, 142,
        143
Late wood..............44, 59, 82
    measuring.............. 96
    relation to strength......47–49
Limit of elasticity........... 5
Limnoria................... 67
Live load..............‥..... 28
Load, application of.......... 29
    concentrated............. 28
    dead.................28, 144
    immediate............... 28
    kinds of................. 26
    live.................... 28
    maximum............... 29
    permanent............... 28
    safe.................... 29
    uniform................. 26
Loading, centre...........97, 100
    static.................. 29
    sudden................. 29
    third-point.............. 97
    vibratory............... 29
Locust..................... 22
    black..........13, 40, 44, 51

PAGE

Locust, honey, 9, 13, 16, 20, 27, 32,
    40, 42, 56
Log of tests, 97–98, 100–102, 103–104,
    107, 110, 114

Machine for static tests......90–92
Maple.................22, 44, 51
    red....13, 20, 27, 32, 40, 42, 56
    silver................... 13
    sugar, 9, 13, 16, 20, 27, 32, 40,
    42, 56
Marking test specimens, 94–95, 100,
    129–131
Material for tests, 88–90, 94, 99–100,
    102, 106, 107–110, 113, 115–116,
    117, 118, 119–120, 121, 122,
    123–125, 128–134, 135, 136
Mechanical properties, definition of, 1
    factors affecting........43–87
Medullary rays (see Rays)
Mistletoe.................. 70
Modulus of elasticity......6, 25, 89
    formulæ, 98, 102, 104, 114,
    123
    of rupture........26, 54, 62, 75
    formulæ..........98, 102
    speed-strength........... 94
Moisture, determination.90, 133–134
    effect of, 6, 8, 17, 33, 75–84, 138–
    139
Mulberry.................44, 51

Natural shape and size....... 2
Neutral axis................ 23
    plane................23, 33

Oak, 15, 22, 43, 48, 49, 55, 60, 66, 71,
    84
    black................... 40
    bur.................... 13
    live.................... 22
    post, 9, 13, 16, 20, 27, 32, 40, 42, 56
    red, 9, 13, 16, 20, 27, 32, 40, 42,
    56, 124
    southern..............52, 71

PAGE

Oak, swamp white, 9, 16, 20, 27, 32, 42, 56

tanbark................ 56

white, 13, 16, 20, 27, 32, 40, 42, 54, 56, 72, 144

yellow, 9, 13, 16, 20, 27, 32, 42, 56

Osage orange, 13, 16, 27, 32, 40, 51, 56

Oven-dry.................. 57

Packing boxes, testing...... 124

Permanent set............'.. 5

Permeability................ 54

Pitch pockets............... 65

Pith rays (see Rays)

Pine...................44, 45, 55

Cuban................. 71

loblolly, 36, 48, 51, 54, 71, 85, 140, 141, 142, 143

lodgepole, 13, 16, 20, 27, 32, 40, 42, 57

longleaf, 6, 7, 8, 9, 13, 14, 16, 20, 27, 32, 36, 40, 42, 46, 57, 59, 65, 71, 78, 140, 141, 142, 143, 144

northern yellow.......... 22

Norway (see Red pine)

red, 9, 13, 16, 20, 27, 32, 36, 40, 42, 48, 57, 140, 141, 142, 143, 144

shortleaf, 13, 27, 36, 48, 57, 71, 140, 141, 142, 143, 144

southern yellow........22, 124

sugar, 9, 13, 16, 20, 27, 32, 42, 57

western yellow, 9, 13, 16, 20, 27, 32, 40, 42, 57

white, 13, 15, 16, 20, 22, 27, 32, 40, 42, 51, 57, 144

Plane, neutral...............23, 33

Plasticity.................. 6

Pliability................5, 38, 85

Poplar.....................22, 44

yellow.................. 44

Pores..................... 44

Preservatives, effect of........ 86

PAGE

Rays...................... 60

effect on compression failure, 17, 18

shrinkage...............81–82

Redwood, 13, 16, 27, 36, 48, 57, 140, 141, 142, 143, 144

Resilience.................2, 5, 49

elastic................... 6

formulæ for.98, 102, 104, 114

Resin, effect of.............. 59

pockets.................. 65

Rind-gall.................. 65

Ring, annual............... 44

growth...............44, 52

-shake...............64–65

-porous...............44, 59

Rot......................68, 69

Rupture, modulus of...26, 54, 62, 75

formulæ..............98, 102

Safe load.................. 29

Sap......................73, 74

-stain.................. 59

-wood...........50–54, 73, 74

Sassafras.................. 51

Season checks............61, 78–84

of cutting, effect of.......73–75

Seasoning............55, 74, 75–84

Second-growth.............49, 50

Set....................... 5

Shake..................64–66, 72

cup.................... 64

heart.................. 64

ring................... 64

star................... 64

Shear................3, 19–22, 133

across the grain.....19, 21, 22

along the grain 19, 21, 76, 94, 107–110

formulæ..........98, 102

horizontal.............. 24

failure.............35–37

longitudinal.............21, 24

oblique.................21, 22

transverse.............. 23

PAGE

Shear, vertical............... 23
Shearing strength............2, 19
Shipping dry................ 57
Shipworms.................. 67
Shock.....................30, 49
Shortening..................3, 33
Shrinkage.... 54, 58, 74, 76, 78–82, 135–137
S-irons..................... 83
Site, effect on wood....48, 49, 70–73
Size of test specimens, effect of, 89–90, 138–139
Sketching test specimens, 94–95, 100, 102, 106
Softwood.................44, 60
Span...................... 25
Specific gravity.......55, 135–137
Speed of testing machine.....93–94
    -strength modulus........ 94
Sphœroma.................. 67
Spike-pulling test............ 123
Spiral grain................. 60
Splintering tension failure...... 34
Splitting..................41, 60
Spring wood (see Early wood)
Spruce, 14, 15, 22, 44, 59, 75, 84, 144
    Engelmann, 13, 16, 20, 27, 32, 40, 42, 57
    red.......7, 13, 27, 57, 140, 141
    white.......13, 27, 57, 140, 141
Static tests, machine for......90–92
Steaming, effect of..........85, 87
Stiffness....1, 4, 5, 6, 25, 26, 62, 76
Strain, definition of.......... 2
    unit................ 3
Stress, compressive.......... 3
    definition of.............. 2
    due to impact...........31, 32
    external.................2, 33
    internal................ 2
    shearing.................3, 21
    tensile.................3, 62
    torsional................ 38

Stress, unit................. 3
    -strain diagram....3, 97–98, 100
Structural timbers, strength of, 138–144
Summer wood (see Late wood)
Sycamore, 9, 13, 16, 20, 27, 32, 42, 57, 64

TAMARACK, 9, 13, 16, 20, 27, 32, 36, 40, 42, 48, 57, 140, 141, 142, 143, 144
Temperature, effect of........84–86
Tensile strength........1, 7, 23, 78
    parallel to grain...........7, 8
    right angles to grain....... 8
Tension.................... 7
    failures................. 34
    tests...............118–122
Teredo..................... 67
Tests, impact...............31–33
    timber................88–136
Test specimens, size of....... 89–90
Timber testing............ 88–136

VEHICLE woods, testing 124
Variability of wood........1, 2, 43

WALNUT, black............... 22
    common................. 22
Water content.......... 55, 73, 74
    effect of..... 6, 8, 17, 33, 75–84
Wear, resistance to (see Abrasion)
Weight, relation to mechanical
    properties............... 54–55
Willow.................... 44
    black.................. 13
Work (see Resilience)....... 30, 54
Working plan 88, 127–137

XYLOTRYA................. 67

YELLOW POPLAR............ 44

ZINC CHLORIDE, effect of....... 87

Also from Benediction Books ...
Wandering Between Two Worlds: Essays on Faith and Art
Anita Mathias
Benediction Books, 2007
152 pages
ISBN: 0955373700

Available from www.amazon.com, www.amazon.co.uk

In these wide-ranging lyrical essays, Anita Mathias writes, in lush, lovely prose, of her naughty Catholic childhood in Jamshedpur, India; her large, eccentric family in Mangalore, a sea-coast town converted by the Portuguese in the sixteenth century; her rebellion and atheism as a teenager in her Himalayan boarding school, run by German missionary nuns, St. Mary's Convent, Nainital; and her abrupt religious conversion after which she entered Mother Teresa's convent in Calcutta as a novice. Later rich, elegant essays explore the dualities of her life as a writer, mother, and Christian in the United States-- Domesticity and Art, Writing and Prayer, and the experience of being "an alien and stranger" as an immigrant in America, sensing the need for roots.

About the Author

Anita Mathias is the author of *Wandering Between Two Worlds: Essays on Faith and Art.* She has a B.A. and M.A. in English from Somerville College, Oxford University, and an M.A. in Creative Writing from the Ohio State University, USA. Anita won a National Endowment of the Arts fellowship in Creative Nonfiction in 1997. She lives in Oxford, England with her husband, Roy, and her daughters, Zoe and Irene.

Visit Anita's website
   http://www.anitamathias.com,
and Anita's blog
   http://dreamingbeneaththespires.blogspot.com, (Dreaming Beneath the Spires).

The Church That Had Too Much
Anita Mathias
Benediction Books, 2010
52 pages
ISBN: 9781849026567

Available from www.amazon.com, www.amazon.co.uk

The Church That Had Too Much was very well-intentioned. She wanted to love God, she wanted to love people, but she was both hampered by her muchness and the abundance of her posses-sions, and beset by ambition, power struggles and snobbery. Read about the surprising way The Church That Had Too Much began to resolve her problems in this deceptively simple and en-chanting fable.

About the Author

Anita Mathias is the author of *Wandering Between Two Worlds: Essays on Faith and Art.* She has a B.A. and M.A. in English from Somerville College, Oxford University, and an M.A. in Creative Writing from the Ohio State University, USA. Anita won a National Endowment of the Arts fellowship in Creative Nonfiction in 1997. She lives in Oxford, England with her hus-band, Roy, and her daughters, Zoe and Irene.

Visit Anita's website
    http://www.anitamathias.com,
and Anita's blog
    http://dreamingbeneaththespires.blogspot.com (Dreaming Beneath the Spires).

www.ingramcontent.com/pod-product-compliance
Lightning Source LLC
Chambersburg PA
CBHW021426180326
41458CB00001B/150